MANUAL FOR A SELF-SUFFICENT LIFESTYLE
FROM HOMESTEAD TO FIELD

AUDREY LEVATINO

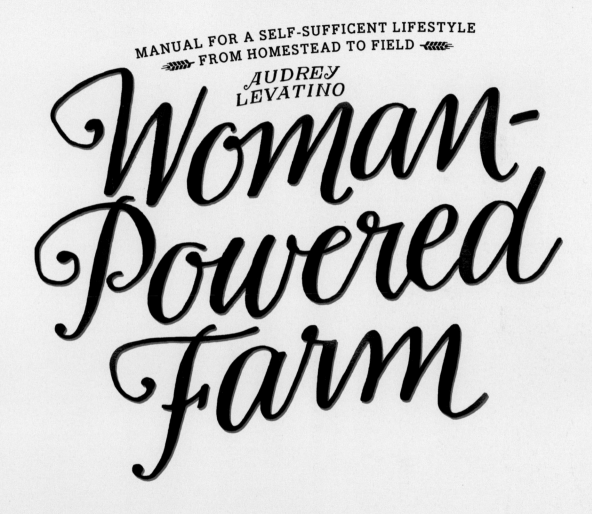

Woman-Powered Farm

COUNTRYMAN PRESS
NEW YORK, NY

THE COUNTRYMAN PRESS
Woodstock, Vermont
www.countrymanpress.com

A division of W. W. Norton & Company, Inc.,
500 Fifth Avenue, New York, NY 10110
www.wwnorton.com

For information about special discounts or bulk purchases, please contact W.W. Norton Special Sales at specialsales@wwnorton.com or 800-233-4830.

Library of Congress Cataloging-in-Publication Data are available.

Woman-Powered Farm
978-1-58157-241-4

PHOTOGRAPHS
All photography by Michael Levatino unless otherwise noted.

Page 9: Alessandro Surai/thenounproject.com; 18: HultonArchive/istock.com; 70: Amy Nicole/thenounproject.com; 75: AleksandarNakic/istock.com; 85: IvonneW/istock.com; 86: Waleed Al-Alami/thenounproject.com; 100: ImagineGold/istock.com; 101: psceric/istock.com; 106: PJ Souders/thenounproject.com; 132: meltonmedia/istock.com; 152: Marco Galtarossa/thenounproject.com; 157: mtreasure/istock.com; 164: NoDerog/istock.com; 178-179: Chart courtesy of Johnny's Selected Seeds, Johnnyseeds.com, 1-877-564-6697; 201: MentalArt/istock.com; 211: manfredxy/istock.com; 215: cjp/istock.com; 220: Adam Zubin/thenounproject.com; 269: Kolbz/istock.com; 270: edelmar/istock.com; 272: pmayne/istock.com; 276: Gabrielle G./thenounproject.com; 318: Diego Naive/thenounproject.com; 322: Tracy Caruso; 329: Anna Caruso

BOOK DESIGN AND LAYOUT
Nick Caruso Design

Printed in United States
10 9 8 7 6 5 4 3 2 1

DEDICATION

To Michael

CONTENTS

CHAPTER ONE
A Call to Farm—A Farm History of Women   16

CHAPTER TWO
Learning About Farming—Internships, Classes,
Resources, Neighbors, and Associations   24

CHAPTER THREE
Finding a Farm—Buying, Leasing, and Inheriting   36

CHAPTER FOUR
The Healthy Farmer—Keeping Your Mind, Body,
and Spirit Safe, Healthy, and Happy   70

CHAPTER FIVE
Getting to Know Your Farm from House to Field   86

CHAPTER SIX
Tools and Techniques   106

CHAPTER SEVEN
Growing on the Farm   152

CHAPTER EIGHT
Farm Animals   220

CHAPTER NINE
The Farm as a Business   276

CHAPTER TEN
Farm Education and Farm Schooling   318

Introduction   8

....

Resources   331
Index   336

INTRODUCTION

Look around your local farmers' market and the first thing you'll notice is how many women are running the booths. Many are also the farmers who tended the goods they're selling, whether it's vegetables, flowers, eggs, or goat cheese. Visit any of your local farms that focus on organic growing methods and sustainability and it is likely that a solid majority of workers and interns on those farms are young women. But not just the smaller and sustainable farms are being run by women. Women are inheriting large farms in record numbers, and running large cattle and commodity operations. Women are noticeably visible as farm veterinarians, agricultural extension agents, and even as managers on the floors of meat processing plants. The face of farming, at least at the local level, has become decidedly female.

Why, after all the years of women (and men) moving off farms and into cities, is there a sudden return to the rural and agricultural work? Women worked for years to get off the farm and away from the constant labor of farm living. Before electricity and in the not-too-distant past, women were expected to help grow the food, make and wash the clothes, cook, clean, and do all the chores of child-rearing, all without the right to vote or the right to inherit the land.

War, a distinctly male pursuit, gave women more opportunities, first in the garden, but then, more important, off the farm. Women in the United States and Europe used their skills during WWI and WWII to raise food for the home front by developing Victory Gardens. Soon, however, there was need for more workers in other areas, so women learned skilled trades, formerly

open only to men, and began working in factories, building planes and bombs while the men were away at war. Women continued advancing the industrial revolution that brought liberation in the form of dishwashers, washing machines, prepackaged foods, and other time-saving innovations. Technological advances also provide automated ways of completing farm chores. Farms that used such machinery required fewer actual farmers. Suddenly, women had more time in their days and could consider leaving the toil of the farm to pursue education and regular jobs. Women had left the farm for personal freedom and have made great strides since then in the business and academic worlds.

Far more women are in the workforce than ever before, but also growing numbers of women are leaving business to pursue agriculture. In the recent farm census, the percentage of woman-operated farms held steady at about 13½ percent. The overall number of farms continues to drop, which has been an ongoing problem. But these numbers overlook a major trend of women incorporating into their lives farming and the growing of their own food. And many of the farm businesses that women are engaged in—flowers, herbal products, artisan cheeses, plant starts, and farm crafts—may not be reflected in a farm census. There's also a large group of community gardeners, urban farmers, and nonprofit farms that do not report to the census as farms, although they are certainly farming food to feed people. A recent article in the *New York Times* tracked down these farmers in the city and reported that these organizations, both the management and labor, are dominated by women.[1] Many women in these agricultural and entrepreneurial pursuits think of themselves as "growers" or artisans in the local food community. Many are raising children as their primary pursuit, meanwhile growing produce on the side to feed their family and create a second income. They don't call themselves farmers, but they are certainly reshaping what farming is and how it affects local communities.

This book does not attempt to restrict the term *woman farmer* to only those women running and earning 100 percent of their living on a traditional farm. Women farmers are now as diverse a group as the country as a whole. They are becoming more diverse in a racial sense—the share of black, Asian, and Native American women farmers (who responded to the census), grew by more than 13 percent from 2007 to 2012—but also in their pursuits. From the high school and college students spending their summer breaks interning on local farms, to mothers growing food for their families and selling the excess at the farmers' market, to single women herbalists growing their own ingredients, to volunteers growing food for the poor, to older women who've inherited land from their husband and are now keeping cattle to protect their agricultural heritage, to artisan cheese makers that keep their own goats and sheep for the milk, the range of pursuits, ages, and experiences is as varied as women themselves.

The term *artisan* has been applied to many of the new, organic, and small producers of food that serve local communities and restaurants. And farming could be thought of as an art. Sure, there are artists who are making their living as painters and musicians. But most artists pursue their craft as a hobby or as a side income and most certainly as a lifestyle. You don't have to make a living at painting, sculpting, or playing music to be considered an artist. In the same sense, making a living should not be a requirement for being considered a farmer. And it's women who are leading the way as artisans of agriculture.

If your income doesn't depend on it completely, farming can offer a great deal of *freedom*. It offers freedom from office work hours, freedom from commuting, freedom to work at home, freedom to spend time with your children, freedom from industrialized food, and all the freedoms of entrepreneurship. And on a deeper level, farming satisfies a woman's desire to nurture and to grow. It offers a way to become more fully a part of one's community, the opposite of being sequestered away in an office for eight hours a day. Farming also offers women *empowerment*. For too long, women have given up control over the food they feed their children to multinational food conglomerates and chemical companies. Farming empowers women to take back responsibility for their own health and that of their families.

Women approach farming differently than men do, both emotionally and physically. This book is written from a woman's point of view and deals with issues that are unique to women. In these pages, you'll find stories of women who have pursued farming—women who've interned or worked on a farm, started their own farm, or inherited a working farm to sustain. But you'll also find practical information about finding and starting and running your own farm business, finding an internship, leasing land, or passing along your farm to your descendants, among other things. There's also practical how-to information about the basic skills that are most valuable to running a farm or country homestead.

Farming became my passion, a labor of love, and the most meaningful work I know to exist. Not only does it take perseverance, but an innate intimacy with the land, plants, weather, and nature itself. Early to rise, often before the sun, a woman farmer cultivates devotion to the work of producing food—good, honest work. As I weed beds, I often feel that I am also weeding my heart and my soul. As I rhythmically plant seeds, the meditation of growing nourishment through organic vegetables enlivens my being. My sincere desire to know that my children and my community are eating organic, local vegetables that I have taken part in growing, gives me a sense of satisfaction.

–ELIZABETH BLECHA
the author's sister and a farmer in Willits, California

Within these pages, you'll find:

- ➧ Step-by-step instructions on chopping wood, operating a chainsaw, basic small-engine maintenance, fence building and repair, and other traditionally male pursuits that every woman farmer should learn, most with step-by-step photos of women demonstrating these skills

- ➧ Instructions and ideas for growing such plants as flowers and vegetables

- ➧ Guidance on taking care of animals, both big and small

- ➧ Advice on keeping your body, mind, and soul healthy for the demanding work that farming requires

My own path fits nicely into the modern journey most women take into farming. My husband, Michael, and I bought our farm together 12 years ago. We cosigned our mortgage together. We shared labor and quickly fell into the traditional roles of the male operating all the power equipment and the woman gardening and taking care of house chores. We both maintained off-farm jobs, he in publishing and me as a teacher.

When it became clear that I was ready for a break from teaching, we made the decision to turn our place into a productive farm that could spin off a second income. My heart led me to pursue growing cut flowers as a business. I attended flower conferences, read all the books, talked to other local growers, and learned by trial and error. As the business has grown over the last seven years, I've taken on the role of principal farm operator.

Michael's job now demands more of his time and he travels often. This has forced me to learn all the chores that were previously his. I've done it all, from burying dead chickens, to chopping wood, to rounding up loose donkeys in his absence. Much, if not most, of the time, I'm a one-woman farm and farm business.

Sometimes living and working alone on the farm can be overwhelming. But it never fails to be both rewarding and empowering. And by all measure, my situation as a self-sufficient woman living on a farm is no longer a novel one. Thousands of women are operating farms, many much larger than my own, and many others are living and raising children in the country in a sustainable way. They all are helping to reassert women's primary place in our food system and our environment. This can only be a positive and healthy development.

I would have liked to have a book like this before jumping into agriculture, and even now, I expect to look back on the interviews I conducted and some of the how-to techniques I've written about. Sometimes reading about another woman's experiences can be invaluable both as support and as useful information.

I hope you'll see this book as a primer and a jumping-off point for whatever might be your interest in farming, whether it be as a backyard gardener, homesteader, intern, small farm operator, or market farmer. I hope that you'll gain inspiration and valuable practical skills so that you can help to continue to transform farming into a more natural, sustainable, and compassionate pursuit that will continue to give women freedom and comfort. Use it to begin weeding your own heart and soul and that of your community.

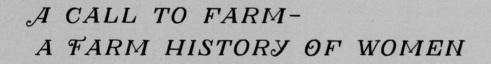

A CALL TO FARM—
A FARM HISTORY OF WOMEN

As long as there have been farms and homesteads, there have been women farmers. It's only natural, when considering the endeavor of farming or even a weekend homestead, to think about the rural women who came before us. What did they grow, why did they farm, and how did they farm? While we want to break new ground (figuratively and literally), we also want a path to follow. For many of us, it is the kitchen garden that lures us into our first taste of country living. We want those fresh-from-the-vine tomatoes and the sugar snap peas that don't even make it to the kitchen, they're so sweet and crunchy. We want to pick our own organic baby spinach instead of buying it at the store. And mostly, we want to reap the bounty of our own efforts in a very elemental way—cultivating and caring for our land and nourishing our body and soul. This is both the lure and the reward of country living.

Security is mostly a superstition. It does not exist in nature, nor do the children of men as a whole experience it. Avoiding danger is no safer in the long run than out-right exposure. Life is either a daring adventure, or nothing.

—*HELEN KELLER*
The Open Door

WOMAN-POWERED FARM

Historically, many women farmed out of pure necessity. Native American women cleared the land, planted the garden, and harvested the crops because men had the jobs of hunting, trading, and going to war. This division of labor dates back thousands of years. The tradition continues to this day, as the second-largest group of women owners and operators of farms and the largest group of minority-owned farms are American Indian or Alaska Native. Ironically, the assignment of primary responsibility for land rights to women in many Native American cultures remains one of the most progressive ideas in rural America today.

Did young men work in the fields? (laughing heartily) Certainly not! The young men should be off hunting, or on a war party; and youths not yet young men should be out guarding the horses . . . also they spent a great deal of time dressing up to be seen by the village maidens . . . But old men, too old to go to war, went out into the fields and helped their wives.

–BUFFALO-BIRD WOMAN
Native American Gardening

Native American women were permitted land ownership before colonists started farming the lands. During the Civil War, it wasn't uncommon for every husband and brother on a farm to be killed in the war, leaving only women behind. These farms were ultimately taken from the women and given over to other men to oversee. But not before the women had put in so much of the work to keep them running.

Following the Civil War this type of shift came during wartime, as men left to fight for their country, leaving behind on the farm households of women, children, the elderly, and the infirm.

ESTHER LEWIS

Esther Lewis, a Pennsylvania farm widow who ran her farm during the 16 years from 1825 to 1841 . . . found ore in her land several years after her husband died, established a successful mining operation, then set up a partnership with a farm tenant, and with other neighbor women produced several hundred pounds of butter for sale each year. Along with the butter, the farm sold poultry, vegetables, grain, and meat in Philadelphia. Lewis educated four girls in the best Quaker boarding schools of the period. In addition, she was active in political reform. She opened her farm to help black families fleeing from slavery, joined early antislavery organizations, circulated petitions opposing slavery and endorsing temperance, and actively supported the women's rights movement. Her farm and the farms of other Quaker women became the seedbed for women's rights.

—Promise to the Land, by Joan M. Jensen

During World Wars I and II, stories of women actively involved in horticulture and agriculture are well documented. The governments of Britain and the United States used the propaganda of patriotism to recruit women to work on farms in "A Call to Farms," and to raise food in whatever city spaces they could find. These Victory Gardens of the Women's Land Army did more than help nations in need of food and workers, they motivated generations of women to take risks, try something new, and push the boundaries of education and occupation that had been conscribed by men. In fact, these wartime efforts of women to grow crops and raise livestock were a significant stepping-stone to women's obtaining the rights to vote and to own property.[2]

From Victory Gardens to today's women-run farms, land has often found its way into the care of women in the absence of men to work it. Women live longer than men, so women often survive their husband, and are left behind to keep the farm going, or at least scrabble enough out of the property to keep themselves and their households fed and sheltered. University researchers have estimated more than 200 million acres of farmland in the United States will change hands by 2027, with women potentially owning a majority of the land. [3]

Women of the last century preserved their knowledge through gardening and unconsciously (or maybe consciously) bided their time. Only since 1982 have women enjoyed equal inheritance rights to land when their spouse died.[4] Recently this struggle has evolved to include changing the way we practice farming, as well as the way our society views farms and farming. Women now lead the way in sustainable and humane agricultural practices, conservation of the land they own, and returning farming to its roots in the community. Women-operated farms tend to be diverse, whereas men lead the way in grain and cattle farming.

The value of farmwork is obvious and tangible. What is not always so obvious is who is doing that farmwork. Up until the 2002 Census of Agriculture, data were collected on only one principal operator (person in charge of day-to-day decisions) of a farm, and this person was usually a male. This approach overlooked the many instances of farms that were comanaged, often by husband and wife. In 2007 the National Agricultural Statistics Service further improved its data collection to include small, minority, female, and limited-resource farm operators. As these new data were compiled, it became clear that more and more women own and operate farms in the United States than was previously known. Although farming is stereotypically considered a male profession, the 2012 farm census shows that women run about 14 percent of the nation's farms, and this is thought to be a big underestimate, as many small farms are undercounted. Women are claiming a larger place in agriculture, through all types of farming and in many different ways.

TEMPLE GRANDIN

Professor of Animal Science and Writer
Colorado State University

Temple Grandin is a professor of animal science at Colorado State University and the best-selling author of *Animals in Translation* and *My Life in Pictures*, among other books.

The best advice that Grandin gives to women (or anyone) wanting to break into farming is, "Get out and see stuff. Get active—network. Go to workshops. If you travel to attend a wedding or funeral, stay over and get out to farms for a visit or visit the local farmers' market." She should know. Over 40 years ago, when she was trying to break into the male-dominated world of animal science, she had to be very persistent to get onto farms and feedlots where a woman rarely ventured. It was this real-world experience that has led to her designing humane animal-handling facilities. And now over one-third of the animals processed in US meat plants are handled in systems designed by Grandin.

She overcame the double obstacle of being a woman and also a person with autism. There were no women out in the feedlots or working on the farms unless they were in the office, answering phones or doing paperwork. It was the fact of being a woman, and not Grandin's autism, that was the biggest obstacle to breaking into agriculture.

"Being a woman was way worse [than being autistic]. When I was starting out, I was a woman stomping into a man's world, and that wasn't easy," says Grandin. And about that scene in the HBO movie about her life where she finds a bucket of castrated bull testicles dumped on her car after working on a feedlot, "That really happened." But it wasn't always the men that were her obstacle. "The cowboys' wives didn't want a young woman out on the farm with their husbands either."

But a lot has changed in agriculture, thanks to pioneers like Temple Grandin. Women and men now attend in equal numbers the animal science classes she teaches at Colorado State University. And the room of "gray hairs" she used to speak to all over the country at cattlemen and other agricultural organizations are now filling up with young people. She even sees women and young people at the meatpacking plants, taking on leadership roles, such as quality assurance.

"Women tend to be gentler with animals. When you see these horrible videos of people abusing animals, I don't think there's one where there's a woman doing the abusing. The really awful things I saw back in the '80s and '90s while doing the McDonald's audits were not being done by women," she says.

And it is women that she now sees populating the organic agriculture conferences and meetings. They seem to be heading up the smaller, specialty operations like goat dairies. They are the ones leading the charge in animal activism.

With so much going right for women in agriculture, Grandin is now most concerned with kids and food security. In her travels around the country she meets children who don't appear to be learning anything practical.

"I've met so many kids that have never cooked, sewed or even made anything out of cardboard," she says. "I'm worried about kids not being involved in real stuff."

She urges farmers to search out children and educate them with farm tours. She feels that farmers' markets ought to encourage farm visits and farms should be open almost all the time. Kids are too removed from the source of their food. This disconnection to the food system is also a threat to food security, according to Grandin, and we need more young people getting into farming to help out.

"What happens when the interstate floods like it did recently out here in Colorado?" she asks. The trucks can't get food to people and it's the local farms that will provide it. "We need Big Ag to help us feed the poor and Little Ag to give us food security."

According to Grandin, internships, such organizations as 4-H, and opening up farms to educational opportunities are the keys to building the next generation of farmers that will be even more diverse than the current one filling up with women. "Kids are hungry to learn about agriculture," she says.

And who better to teach them than the women who are returning to farming today?

LEARNING ABOUT FARMING—INTERNSHIPS, CLASSES, RESOURCES, NEIGHBORS, AND ASSOCIATIONS

CHAPTER  TWO

If seed catalogs have become your pleasure reading of choice, and you find yourself puzzling over the bewildering yet intoxicating number of varieties of tomato seed that are available, you can be fairly certain you've caught the "farming bug." And in its initial stages, the feverlike, heady enthusiasm can be as all consuming as any new addiction (think: shoes, or kitchen gadgets, or yarn). One all-purpose cucumber? You must be joking! There are cucumbers for pickling, for salads, for sandwiches, and for snacking, and what about that intriguing "lemon" variety? You must have them all. Each page in the catalog is an opportunity for indulgence. This delight in the possibilities of growing things and raising animals never goes away. Fast on the heels of the dream of achieving a sustainable country life is the reality of labor and supplies, and the necessity of planning and resources.

The value of doing something does not lie in the ease or difficulty, the probability or improbability of its achievement, but in the vision, the plan, the determination and the perseverance, the effort and the struggle which go into the project. Life is enriched by aspiration and effort, rather than by acquisition and accumulation.

–*HELEN NEARING*
The Good Life: Helen and Scott Nearing's Sixty Years of Self-Sufficient Living

---

All of us want to learn more about this notion that has captivated us, of living on and with the land. To succeed and flourish we need to learn all we can about resources, methods, and plans of action. We want to know how others went about getting started and how they continued on when they got stuck.

The Internet has become a rich resource for farmers. You can now find articles from dozens of agricultural universities at the press of a button. Countless videos on YouTube demonstrate all kinds of techniques that you might need. The resources and forums available are very helpful and rich indeed. I find myself searching the forum for the Association of Specialty Cut Flower Growers almost weekly, for tips in dealing with one problem or another.

I have always been a devoted patron of the library. Because I am drawn to the tactile and the visual, I love to look through actual books, mark pages with sticky notes, and keep a log of what I've discovered. Many wonderful books cover a myriad of farming topics; see Resources (page 331) for a list of titles, including many recommended by many fellow women farmers.

Classes and workshops are not only a great source of information, but also a wonderful way to meet people who share similar interests and to gain exposure to new ideas. We are fortunate to live in a time when access to agricultural classes is readily available through a variety of means. The local community college, environment council, university extension cooperative, and local farms provide a diverse offering of classes from growing mushrooms to beekeeping, whole living, and herbal health and safety, and the list goes on. If you want to learn about it, the resources are out there. Often the best resource is other farmers in your community. The value of shared ideas and experience, getting advice from someone who has "been there and done that," cannot be overstated.

Of course, nothing compares to actually getting out and doing. Choosing a project and working through it from start to finish is a wonderful way to gain knowledge and feel a sense of

Some of the best farm education comes from attending farm classes like this one on the benefits of fungi in the garden at Sharondale Farm, Cismont, VA.

accomplishment. Whether it is building your first chicken coop, planting your first kitchen garden, or deciding to devote six months to interning at a local farm, digging in is the only way to know whether you will enjoy it.

Internship or Apprenticeship as a Path to Farming

You don't need to jump headlong into farm ownership to learn about farming. Many women find it more prudent to learn the skills they will need before looking to buy a place of their own. Internships and apprenticeships are good ways to kick the tires of farming. They are also good ways to experience the less glamorous side of farming and learn which type of farming you don't want to do.

Most interns are expected to do the grunt work. But that's exactly where you need to begin. If you can handle double-digging beds, wrenching wire grass out of pathways, and spreading tons of mulch, then you know you can handle and appreciate the farm tasks that need more skill. Apprentice-ships are a more advanced form of internship. You learn more farming skills and management, but an apprenticeship requires that you already have some basic farming skills.

Be very careful in choosing a farm in which to work. Some farms do not follow all the national and local labor laws and may offer unpaid internships in violation of those laws. You could also find yourself in a manual labor situation that offers little opportunity for education.

But some of the best people you will ever befriend will be the kindred spirits you learn from on farms. And while there are known labor issues with the Big Ag farms, the local, sustainable, and organic farms are generally run by kind people. A good place to seek out internships and apprenticeships is with the National Sustainable Agriculture Information Service. This site offers hundreds of farm internships listed by state along with lots of useful information for farmers of any level. However, be aware that the service does not screen or endorse the job postings or farms on its website. That said, the listings are very informative and will answer your questions about pay, living conditions, and what you might expect to learn as an intern or apprentice.

ANGEL SHOCKLEY

Herb Angel
Charlottesville, Virginia

Angel Shockley's love of plants and herbs began in college. It was common houseplants that sparked her interest. She was most interested in the medicinal qualities of plants, so she started to search for this knowledge by interning on several farms around the country. As a young woman fresh out of college, she was able to move around and live simply without much money, while collecting the knowledge she needed for her next step in agriculture or whatever else life brought her way.

Her first intern experience was working for an eco-tourism farm in California. This led to wanting more actual plant farming experience. She moved on to a 5-acre farm run by just one single woman in Montana. It was hard work, and she realized that it was hard to make a living at farming. Some of the interns left because of the difficulty of the work. On a 5-acre production farm that required serious grunt work every day, she didn't have the time or opportunity to acquire the experience that she desired with the actual plants. She quickly realized that plant farm production on a large scale prevented her from developing an intimate relationship with the plants, which she now knew was what she most wanted out of farmwork.

But it wasn't all toil and no fun. She developed close bonds with the other interns, and the owner made efforts to help educate them about the work. There was very little money, as with most internships—$20/week, compensated by food and lodging. Also, they were isolated from town/city life, which left little in the way of social opportunities for a young 20-something. Angel stayed for one season before realizing that she needed to rethink how to make a living working with plants. Often, what is most valuable to learn about an intern experience is what kind of work does *not* suit you.

She returned to Virginia, where her family lives, and quickly found another internship, this time on a farm where she might learn more about plants. She found the internship via ATTRA, the website for the National Sustainable Agriculture Information Service. She signed on with a well-known farm in the Charlottesville area and was paid more than at her other internships—$800/month, free vegetables, and free on-site living accommodations. However, the on-site housing proved to be unfinished, without a bathroom, and was very cold in the winter months when she started. But the owner had real farm skills and the farm was close to the town where Angel could have a social life as well. She stayed for a full year, which provided her with a feel for what working on a full-time farm is all about. The low pay, difficult work, and hard living conditions did not detract from her interning experiences.

There was always a clear agreement and I knew what I was getting myself into. Because so many of the children of farmers leave the farm, interns are kind of like the 'new family' to these farmers. What made interning satisfying was the lifestyle—it was challenging and hard work, but the social aspect and doing communal work were most rewarding," says Angel.

What did Angel learn from her time as an intern? That herbs were her calling and that led her to undertake a multiyear education in herbalism that has led to her current business, Herb Angel, selling herbs, teas, tinctures, salves, and other useful and medicinal herbal products. She learned as well that she did not want to be a market farmer; that interning is intensive work that requires a conscious effort to achieve a sustainable balance for body and mind; that she preferred the term *apprentice* to the term *intern*; that all farms that employ interns should use more than one so as to facilitate the shared experiences and bonding that occur in farmwork; and that Charlottesville, Virginia, is where she found her path and her calling. Without the experience of internships, that path may never have revealed itself.

A Selection of Internships and Farm Education:

- **National Sustainable Agriculture Education Service:** Internships on sustainable farms.

- **AFS (formerly American Field Service):** For high school students, AFS is a leader in intercultural learning and offers international exchange programs in more than 40 countries around the world through independent, nonprofit AFS Organizations

- **Good Food Jobs:** Jobs and internships in the sustainable food world, including farming.

- **World Wide Opportunities for Organic Farms (WWOOF):** Woofing is for the more adventurous souls whose idea of a vacation is working on a farm, somewhere around the world. WWOOF links farm volunteers to international organic farms and growers. You can use it for opportunities domestically as well.

- **Ecology Action:** Biointensive Farming has partnerships all around the world, teaching people to grow their own food in a healthy, sustainable way.

- **4-H:** A program of the Cooperative Extension System of Land Grant Universities, 4-H is the nation's largest youth development organization. Although it covers numerous fields, farming is a significant component.

- **Future Farmers of America (FFA):** As its name implies, the FFA helps trains the next generation of youth to feed the world.

- **Center for Agroecology & Sustainable Food Systems (CASFS) at the University of California, Davis:** One of the few formal training grounds in the United States that works to advance food justice.

- **Warren Wilson College, Asheville, NC:** The college integrates extensive work and service programs to complement its academics. These include farming, landscaping, construction, and much more.

As with any job search, you need to make your own judgment call regarding an internship's compatibility to your goals. And remember, farms need interns more than interns need farms. Make sure the situation you sign on for will truly help you in your farm education. One very interesting opportunity to learn some farm skills is through the World Wide Opportunities for Organic Farms. Linking international travel with a working farm vacation, or "Woofing," is the modern version of backpacking across Europe and staying in hostels. You get to experience how people in another part of the world live in a rural setting. It's an authentic means of tourism. However, if domestic internships can be hit or miss, you can bet that international ones are even harder to screen ahead of time. So, be adventurous, but do your homework before committing to a beet farm in Kazakhstan in the winter.

Gail Hobbs-Page of Caromont Farm illustrates the intricate
scheduling that goes into making farm-fresh cheese.

WOMAN-POWERED FARM

How to Find the Perfect Internship/Apprenticeship

- Start with woman-owned/operated farms! After all, they are the fastest growing and most dynamic of all farms. If it's not woman-operated, ask how many women work on the farm and how many you'll be working with directly.

- Match the farm with the skills you want to learn. It seems obvious, but raising goats for meat and raising goats for mohair are two different endeavors.

- Ask to be referred to a former woman intern so that you can ask her about her experience. Also, beware if you're the first intern on a farm. It shouldn't be a deal-breaker, but you may find that the farm has not properly prepared for an intern and the position may be less structured than you'd prefer.

- Don't go to work for a farm that will be your main competitor. It's just not cool to take the knowledge from a farmer and then go into business as his or her direct competition. Best to travel to another town to learn the skills and then bring them back to your home base.

- Look for a farm in the climate zone and terrain that you'll eventually be working in. The skills you learn farming in New Mexico don't translate well to Maine.

- Be clear what the living situation is and make sure you're prepared for it. You could be living in a yurt (and loving it!).

- Visit the farm first. If it's not practical to do that, then work out a one- or two-week trial internship to make sure you and the farm are a good fit.

- Look for internships where you'll be working or living alongside at least one other person (maybe your own partner). It's always valuable to have two minds learning tasks, and having companionship is important for what can be a lonely job.

FINDING A FARM—
BUYING, LEASING, AND
INHERITING

CHAPTER THREE

So, you want to move to the country. Perhaps you've been raising your kids (or just yourself) in the healthiest environment you can muster in the city. You've built a nice kitchen garden off your back porch, snipping herbs for meals, feeding yourself fresh vegetables and berries from carefully pruned bushes, even keeping a couple of chickens for their fresh eggs. You go to the farmers' market almost every week and supplement your own home-grown food with that of local farmers. But now you want to more fully practice what you preach and incorporate farming into your life.

Or maybe you've built a comfortable nest egg by putting in long hours in the corporate world and you're looking to change your lifestyle. Perhaps you inherited some money from a loved one and you're looking to invest it. Preserving and creating a productive farm is a popular and noble way to invest in your own future and that of rural America. Or perhaps you're fresh out of school and the idea of working close to the land or animals, instead of a work life spent indoors, is calling to you. There are many paths to a farm.

My own path to a farm started with a simple desire to move from city to country. I was living with my boyfriend (now husband), Michael, in Oakland, California, during the dot-com boom. We shared a one-bedroom apartment on the top floor of an art deco house and paid $400 per month more than we now pay on our mortgage for 23 acres in Virginia. It was a lovely place to live, but we sorely missed the lack of land. Both of us had been avid gardeners while growing up, and felt the draw of home ownership as well as some space to spread out. So, when we had

WOMAN-POWERED FARM

the opportunity to relocate on the East Coast, where property was more reasonably priced, we decided to give it a shot. Still, we had no idea a farm was in our future.

Our long-distance search began. Michael would fly east, rent a car, and explore the cities and towns in the mid-Atlantic region, looking for a place where we might be happy. He was a sales representative at the time, so we had some options geographically. We wanted to live in a place with bookstores, restaurants, a thriving music scene, and natural beauty. College towns usually have this kind of atmosphere along with a youthful energy, so we decided on Charlottesville. The first meeting with a Realtor was instant reality check. Turned out, we could not afford to live in Charlottesville and own a house. An apartment, sure, but we were ready for a yard and a dog. So, out of town we headed. Not too far, just a few miles.

We looked at maps and searched in a wider radius around town until the prices came down. The farther you get from the center of a town or city, the more you can generally get for your money. Then, after driving hundreds of miles of back roads, viewing dozens of houses, and spending countless hours on Realtor websites, Michael spotted a classified ad in the local paper. It beckoned us to: "Bring your animals and live the country life—fenced for horses." We had no desire to have horses, but knew there must be a good amount of open land if it was fenced for them.

The house needed a whole lot of love, to say the least. I was in Oakland when Michael found it and could only see it through photos. He must have cropped out all the bad stuff. We had to move quickly, but we just weren't sure about owning a farm. Michael's mother, Maryneil, lived on a ranch with her husband, Tom. They gave us the terrific advice that too much land is never a bad thing; you don't have to do anything with the land if you don't want to; you can start small and only farm what you want (or not at all) until you feel comfortable doing more. So, we took the plunge.

For me, the property remained sight unseen as I was still finishing my teaching job on the West Coast. Did I mention that the house needed a lot of improvements and upkeep? Here's how much: According to her own account, Maryneil meanwhile went to Virginia to visit family, so she went by to see the farm. As she and Tom drove up the drive and caught sight of the place, she burst into tears. She wept openly that it would be the end of our relationship and our economic future. Tom made her collect herself before she got out of the car to meet Michael at the house. She pulled herself together and she and Tom set about talking to Michael about how to improve it. Lucky for us, Tom's an architect and Maryneil is an interior designer.

Twelve years later, we're still working on this house. But it's been 12 years of steady improvements. The entire endeavor, from creating our domestic life and living situation together to building my farm business, has grown slowly and organically. If we'd not overlooked the obstacles and looked for the silver lining down the road, we'd never have found an affordable farm to pursue the lifestyle we have learned to love. At the time, we only had an idea of the lifestyle that we wanted for ourselves. So, it took a leap of faith to follow the signs to that life that others may have passed up because it seemed like too much effort to achieve.

There are as many reasons to own or work on a farm as there are farmers. When embarking on your own path, it's important to clearly define your goals.

Questions to Ask Yourself When Buying a Farm

◆ Do you want to farm as a lifestyle?

◆ Do you plan to be a subsistence farmer or homesteader to independently provide for your family's needs?

◆ Are you looking for an investment for your money?

◆ Or do you want to be a professional and pursue farming as a career?

The farm or land you choose should fit your goals—but don't forget that goals change. So, try not to limit your future options.

The spirit to adopt when you begin searching for a farm or taking over the operation of a farm is the same one that British and American women coined when creating the Women's Land Army and the Victory Garden movement that both fed nations and lifted women's work up to the level of national patriotism—they cheered the slogan "Dig for Victory!" You will certainly be doing plenty of digging in the literal sense, but you will also be digging to find your own way of life.

But there will be obstacles, both physical and emotional. Whenever people, especially women, decide on a course that's unconventional, they are met with suspicion of their motives and outright jealousy. You will hear, "Living on a farm is so much work. How will you do it?" or "Farms are not profitable. Don't you know it costs sixty-four dollars to grow a tomato?" There will be endless stories about an uncle, or a second cousin, or friend of a mother-in-law who moved to the country only to dig themselves a grave of toil, or lose a hand in a combine accident or throw away every penny of their savings on a money pit. There's a pervasive idea that anyone, especially a woman, getting into farming is a dreamer, naive to the point of self-destruction. Farming is hard and costly and risky—everyone knows that. But so is being a teacher, bookstore owner, or anyone else in a profession that provides more than monetary rewards.

Clearly defining your goals will help you face doubt from others as well as yourself. And you will certainly encounter your share of self-doubt; that is inevitable when embarking on an uncommon path. Remind yourself that you are a modern pioneer woman. The pioneer women (and men) before you were not born knowing the skills to be self-sufficient. They employed a spirit of discovery to their needs and learned from others with more experience how to take care of themselves in the natural world over time. And you can, too.

This is not to say that the warnings of well-intentioned friends and family do not hold a kernel of truth. There might be as many failed farm *businesses* as successful ones. But the only failed *farm* is the one that's been sold to a developer. Just the act of buying land and preserving it is a victory. You do not need to plant one tomato or raise one sheep to be an essential part of your environment.

It's no secret that farming has been a tough business for many years. As the result of decades of mechanization of our farming, government subsidies that favor the big food corporations and the industrialization of our food system, many small farms that relied on solid, local customer bases to sell their products dried up. National chain grocery stores forced out small independent markets and the food system consolidated, much the same way small bookstores, toy stores, hardware stores, and other independent businesses have been under threat.

But like the 18-year cycle of cicadas that we have here in the mid-Atlantic, every living thing has a cycle. And farms are certainly living things. The economics, and especially the culture, are turning once again to favor the local, and farms are reinventing themselves. Even if you do not make one penny on your farm, you are doing a tremendous service to your community just by caring for it. Rural America needs the money that fled years ago to return and invest.

Women have made great strides in corporate America. While we have not yet reached true pay parity, many of us now have the economic and educational means to reinvest in the land, either with actual money or our unique perspective. This is where you come in, whether you're moving from a job in the city, supporting an established farm through an internship, buying a weekend house in the country, or you've inherited a chunk of land. Whether you have money or not, you as a willing participant are valuable farm capital that will help farms grow and thrive.

The benefits of owning a farm, even if you never make a penny in monetary profit:

- Purchasing (or not selling) land keeps it safe from development and environmental damage.

- Low-density population in rural areas conserves our depleting water supply.

- Growing and eating your own food is healthy and delicious!

- The money you spend on supplies and tools creates jobs and tax revenue.

- Managed correctly, your farm is a giant carbon offset, building soil and protecting trees that remove carbon dioxide and generate oxygen in our atmosphere.

- Protecting wildlife habitat protects biological diversity.

- And most important of all, owning a farm can benefit your quality of life.

CAMEY STEWART

Honey Creek Angus
Fredericksburg, Texas

Having come into farming, or ranching as they call it in Texas, in middle age, Camey Stewart was not quite prepared to be a one-woman rancher, but it was thrust upon her. She'd lived in the city most of her life and only moved to the country when her husband decided to slow down after building a successful mortgage company. He was diagnosed with cancer soon after they moved to Fredericksburg and he died a few years later. Camey was suddenly the woman owner of a ranch and was left with a lot of money.

Luckily, they had a ranch hand, Ruben Sagebiel, who had a good handle on dealing with the small herd of cows on the property. Ruben is straight out of central casting as the strong, silent cowboy. Camey does plenty of talking for the two of them. Sometimes it's not at all clear who the boss is, but deep down you know it's Camey's show and Ruben is happy in his behind-

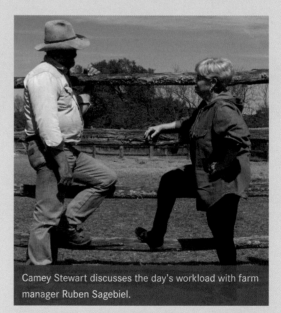

Camey Stewart discusses the day's workload with farm manager Ruben Sagebiel.

the-scenes role as a clear-eyed realist to Camey's exuberance.

"I didn't know what I was going to do with my life [after my husband died]. I'm an only child and had no remaining family to rely on.

But I had a lot of cash. So, I decided to buy a ranch," says Camey. The main reason behind this idea was to have something concrete to pass along to her sons.

She enlisted the help of a Realtor and toured countless properties. The Realtor was somewhat dismayed when Camey fell in love with the roughest piece of property they'd toured—full of invasive cedar and scrub brush. But Camey felt that God had brought her to this piece of land for a reason. She set out immediately to create a natural and beautiful environment that, first and foremost, improved the land.

She made the wise choice to enlist the help of the Natural Resources Conservation Service (NRCS). The NRCS is part of the USDA and provides farmers and ranchers with financial and technical assistance to voluntarily put conservation to work on their lands. The service advances the environment as well as the agricultural well-being of farms and ranches. The NRCS sent representatives to the new ranch and helped her develop a five-year plan to improve her ranch ecosystem.

She hired Ruben to be her permanent ranch manager and now keeps hundreds of heads of cattle on her property. She won an outstanding conservation award from her county and then went on to win the bigger award for her region. She had truly made the land productive and not just a plaything for her children to inherit, which was always her goal.

She made plenty of mistakes along the way, including trusting a "witcher" to find water on her property. A witcher, or dowser, is a person who claims to be able to locate water underground using a forked stick. It's an age-old practice that has never been proven scientifically. She drilled where the witcher suggested three different times, all the while telling the conventional driller that she was certain of the result. But there was no water. Thousands of dollars later and a mile of pipe to bring water in was what it took to overcome this mistake.

Camey has some advice for other women who might find themselves alone with a farm to manage. "Ask for help, because it's free. There are offices and organizations there to help you. They have lifetimes of knowledge, if you just ask for it."

The Farm as Inheritance

If you find yourself in a situation where you've inherited a large farm, you should of course search out the advice of a qualified estate planner who understands generational succession. There are a lot of tax issues to be aware of, including the estate tax. Right now, the threshold for the estate tax is $5.3 million, meaning that any portion of the property you inherit or pass along over that amount is subject to the estate tax. One way that many farmers are ensuring that their farms are passed along to future generations without burdening their heirs with taxes is to place them in a trust over the course of many years.

A LIVING TRUST

A revocable living trust can control all your assets both during your life and after your death. Farmers who set up a living trust can transfer the title of all their major assets (farm real estate, stocks, bonds, etc.) from their names to the name of the trust. They won't lose control of the assets because they would name themselves as both the trustees and beneficiaries of the trust.

Leasing a Farm or Farming a Lease

Before we talk about finding a farm or land for purchase, let's consider a lease, whether you are looking to be the lessee or lessor. It used to be that leasing was mostly for the short term. A farmer would lease some land each season to cut hay or grow crops. Maybe the land would be leased to run cattle for a few months before being sent to a feed yard.

Short-term leases or rentals are certainly an option, if it's your only option. Short-term leases or rentals are least desirable for both parties. While both offer flexibility, neither offers stability. Nor do they foster a connection or responsibility to the land. If you are leasing your land out on a short-term basis, the farmer has no incentive to invest in the soil, pastures, or structures on your farm for the long term. The lessee will be incentivized to get as much out while investing as little as possible. And if you are the one doing the renting/leasing on the short term, whatever work you've put into the farm might be taken away very quickly. Farming is a long-term proposition if you want to have any chance of success.

With land becoming so expensive, long-term leases (those of 20-plus years) are becoming more common. Fueling some of the movement to long-term leasing are the large swaths of land that used to be farmed and are now held in family trusts. The families that inherited the land may not have any desire to continue to farm it. But they would like to keep the land as a working farm to enjoy the tax benefits and also for their own personal desire to preserve the farm heritage of their family's home.

As a woman, you may be looking to find a long-term lease for you to locate your farm business, or you may have a large inherited or purchased farm for investment that you would like to lease to a farmer. The benefits of a long-term lease are becoming more pronounced as land values rise. Farmers just starting out usually have very limited resources and cannot afford the most desirable and fertile land. By leasing, you can be located near urban areas where there are more customers, without paying the exorbitant prices it would cost to own that land. And the money that you save by leasing can be put toward equipment and growing your farm business instead of to property taxes and bank payments. You can gain flexibility and access to bigger markets, while still assuring that you will be

Tasting the fruits of a farm before you purchase is
also suggested!

able to stay on the land for your working lifetime or beyond. And you may be able to trade some of what you produce instead of only paying in cash.

The benefits of a long-term lease for the farm owner are many as well. As mentioned before, you can continue to enjoy the tax benefits of having a farm without doing all the work yourself. You basically have a caretaker of your land that, if the lease is done right, will be improving and preserving your land. You can work out clauses in the lease to ensure that you receive a share of the food produced on your land. And you can support other women by giving them a helping hand in their quest to farm.

University researchers have estimated more than 200 million acres of farmland in the United States will change hands by 2027, with women potentially owning a majority of the land.

–USA TODAY[5]

Long-term leases provide stability and flexibility for both the lessor and the lessee. It's a viable option to help new farmers and also new or current landowners to keep their land in farming. However, you need good lawyers to craft the lease to anticipate all the possible issues that might come up over 20 years or more. The relationship between lessor and lessee in a long-term lease is extremely important; in fact it's the most important requirement of a successful partnership. Expectations have to be open, honest, and mutually agreed upon.

PROFILE OF A LONG-TERM LEASE

At a farm conference in Virginia, a woman farmer shared the story of her successful long-term lease. She and her husband were graduate students living near Washington, DC. They worked in the summer on one of the more established vegetable farms and helped sell what they grew at one of the biggest farmers' markets in the region. The more they worked and learned on the farm, the more they thought they might want to give it a shot as a way of life. The only problem was that land in their area is incredibly expensive. Most of it is either held in big family land trusts or people with big money from Washington, DC, have bought it up as leisure property. Farms are becoming a scarce resource around our nation's capital.

But they were young and willing to start with nothing and work their way up. So, they began living full-time, in something resembling a yurt, on the farm that was employing them. They learned everything they could about farming and worked hard. They became friendly with many people from the farmers' market, both farmers and customers. When a landowner began looking around to find someone to farm his land for him so he could enjoy the all the benefits (taxes, food, and heritage) of having a working farm again on his family land, he began asking around farmers whether they knew of anyone who might be inter-ested. Because they'd networked with farmers in the area and proven themselves to be good growers, they were recommended.

With a little monetary help from their family, they went all in. They moved to the big farm for a two-year trial period, and living in an old picking shed with no heat, in exchange for improvements to the picking shed and food that they grew. They rented farm equipment hourly from the owner. In the meantime, they received another lucky break as a spot opened up at the farmers' market for them to become permanent weekly vendors. This particular market is big and has a waiting list a mile long. But since they'd put in their time there, getting to know the market managers and proving themselves to be competent and dependable, they earned their spot. If they could just produce enough food to sell, they could potentially make thousands of dollars a week at this market and make a living as full-time farmers.

The next two years, they began to pay for the lease in cash and CSA (community supported agriculture) shares. They began exploring ways to create a long-term lease. After months of negotiations with their attorneys, when they spelled out every eventuality, a 20-page, 40-year lease was agreed to, and they've been successful and happy ever since.

Basic Overview of a Farm Long-Term Lease

- The period of lease is 40 years, beginning with one five-year term and seven additional terms.

- The farmers own the buildings and lease the land for cash.

- The owner pays the property tax.

- The farmers can build whatever buildings they want, including a home. The owner must buy back the buildings at a fair market value when the lease terminates.

- The farmers can terminate the lease within 90 days' notice, but the owner must honor the lease during the five-year increments unless there's sufficient cause to terminate early (such as illegal activity or something else egregious).

- Lease stays with the land and automatically transfers to the farmers' descendants and the owner's descendants. So, even if the farm is sold, the lease would have to be honored by the new owner.

- The farmers are allowed to sublet 25 percent of the farm.

- Both farmer and owner agree to not use pesticides and keep the farm as certifiably organic.

- The farmers must keep the land as a working farm, but they are allowed to take three months off a year and to take a sabbatical occasionally.

- The farmers have to maintain their primary residence on the farm and they have to be the farm managers.

Since securing their future on the land while not digging themselves a hole of debt, the farmers have been able to make a living for themselves and start a family. They built a house, are able to put money away every year into a retirement account, and take a vacation every off-season. The long-term lease that they created has allowed them to beat the odds and make a comfortable and stable living as small farmers. They've proven that it can indeed be done.

A Farm of Your Own

When you begin your own farm search, go into it with an open mind. Every farm has its own individual identity. Like all living things, no two are alike. Clearly define your goal for the farm—lifestyle, subsistence, or career. As your goals now may be different later on, balance your present needs with what might come up in the future. Although you may just want to live on a farm for now as a lifestyle, you may find over time that your passion is leading you to farm for a living. Keep your options open.

How to Find a Farm or Land to Buy

➧ **Raised Farmers' markets:** If you plan to operate the farm as a business, first locate the markets where you'll sell your products. Usually that will mean a city or town where there's a concentration of customers. Then search in an ever-widening radius around that central location until you find suitable land at the price you can afford.

➧ **Internet and newspapers:** There are countless sites for finding real estate and land. Realtor.com, Unitedcountry.com, and Craigslist.com are good places to start. And some newspapers still have classified ads. Rural communities usually have a local, free rag that mainly serves this purpose. Look for them near the door of the area's convenience stores and markets.

➧ **Country Realtor:** Look for Realtor signs in the rural areas where you want to live and contact the Realtors that represent properties that appeal to you. Many Realtors specialize in city or country living. Best to try to find a Realtor who has experience buying and selling farms or land.

- ➡ **Farmers:** Go to a farmers' market and ask around whether anyone knows of land or farms that are for sale. Farmers are the first to know when their neighbors are looking to sell.

- ➡ **Feed/supply stores:** Most have bulletin boards with notices of land, animals and equipment for sale.

- ➡ **Take a drive:** Try every back road you can find in the area you are targeting. Many people nowadays are cutting out the real estate middleman and selling the properties themselves. "For Sale by Owner" signs will point the way.

Three key elements to look for in a farm while keeping your goals in mind are the *potential of the land* for specific types of farming, the *location* of the farm, and the adaptability of the land or farm to your *preferred lifestyle*.

Potential of the Land

The potential of the land is the most important consideration if you plan to operate the farm as a business. It's less, but still important if you want to farm as a lifestyle or homestead for your family's needs only. But don't be overwhelmed by trying to define the full details of your life plan when you begin your search for a farm. Gaining some knowledge for the potential of land will only help you define your goals more clearly as you go along.

$SOIL$, or carbon, to be more exact, is one cornerstone of a farm's potential. Take a shovel when you're visiting possible land to buy or lease. Dig out a shovel full of dirt in all the different areas of the farm. Smell it and try to crumble it in your hands. Is it heavy clay, which will make it more difficult (but not impossible) to grow vegetables or flowers? Is it sandy, which offers good drainage, but might not hold enough water for such crops as corn? Or is it nice, dark, and loamy, which would be ideal for any type of farming? Is there a puddle of water in the bottom of the hole, which would indicate you might need to install costly drainage?

Find out what type of farming, if any, has been successful before on the same piece of land. Look for earthworms. Very healthy soil has 10 to 20 in one cubic foot. Another good idea is to take soil samples from areas you can envision using in production or grazing and have it tested. You'll quickly know whether you will need to spend time and money improving the soil content and structure for the type of farming you want to do. It's been said correctly that soil without biology is geology. Or that soil without a thriving living food web of bacteria, fungus, nematodes, and worms is just rock of one form or another. Poor soil quality is not a deal breaker. But you may need to factor in time and money to improve it. And poor soil quality may help you get a better deal on the purchase price.

$WATER$ is the lifeblood of the farm. Look for streams, ponds, and rivers and find their sources—a stream running from an intensive cattle operation next door might be unusable on vegetable crops. Ask whether there are natural springs and where they are located. Ask if the well has ever run dry—then go ask the neighbors, who might have more objective feedback on the local water supply (see chapter 5 for more information on wells). Look for erosion or standing water. Standing water is a bad sign if you want to graze animals and you might need to spend extra money fencing animals off from wet areas or

This red clay soil works well with vegetables because it retains water.

A full pond is a sign of a high water table and a healthy farm.

streams. Check drought and water quality information at the USDA Water Quality Information Center (http://wqic.nal.usda.gov). Determine whether the property is in a flood zone by going to the FEMA website (www.fema.gov).

TOPOGRAPHY is also important. Is the ground sloping or even steep in areas, which will make it hard to plant crops? On the other hand, sloping, steep, and rocky land is not so bad for raising such animals as goats, sheep, llamas, or alpacas. If you plan to grow food or flowers, does the land have good sun exposure and what direction does it come from? Most plants prefer as much sun as possible, unless you're growing certain flowers, such as hydrangea. Southeastern exposure is generally the ideal for vegetables and most flowers. Northern exposure is usually least desirable. But it all depends on your location and the climate. Ask around if you aren't familiar with the growing conditions and find out what works and what doesn't in the topography of the area.

How fertile is the land? *FERTILITY* in relationship to the land is its ability to reproduce crops and provide animals with the ideal environment to reproduce. Is the farm currently producing a good amount of crops and revenue? Is it certified organic or does it have the potential to be certified soon? Is it a monoculture? A diversified farm of rotating crops and animals points to good fertility. Monoculture farming, overgrazing, and sites of trash dumps (more common in the country than you'd imagine) point to a lack of fertility.

Identify the *PLANT COVER* as well. Once you're further along in the process of deciding on a farm, you can ask the agricultural extension agent to meet you out there to help identify the plant cover if you aren't able to do that yourself. Are there problem plants that will require a lot of effort or money to remove, such as large areas of honeysuckle or poison ivy? Is there a field of purple nightshade, deadly to horses, right where you envision your grazing pasture? Will you need to clear a lot of trees and brush to open more fields for planting? Out here in Virginia, we have all of those potential problems. The one most irritating to us is Bermuda, or wire, grass. This terrible weed grows and reproduces with its wiry roots that tangle with other plants' roots and choke them out. Cutting them up with a tiller just helps them reproduce into other smaller plants, like in some alien horror movie. The only way to get rid of it without using extremely dangerous chemicals is to hand dig it out with a shovel. We asked other flower farmers where we live to share their biggest mistakes in farming; most pointed to planting perennials in an area that was full of wire grass without first digging it out.

FARM STRUCTURES are a two-way street. New, sturdy barns, outbuildings, and fences are ideal, but they probably add a pretty penny to the price of the land. Old barns that need work might be aesthetically pleasing to the eye, but you'll need to factor in your ability, time, and economic means to improve them. Fencing especially is important if you plan to have animals. There are specific types of fencing for specific animals and you'll want to know whether your fencing matches your planned farm endeavor.

Location

Settling on a location goes hand in hand with your goals. If you're looking for a farm for the privacy, then living far out of town might suit you. But maybe you don't feel comfortable living 30 miles away from a hospital or grocery store. If you want to have a working farm, you'll need to consider how far away the market for selling your product is. Here are some things to think about when deciding on a location:

- If you or your partner is also working in an off-farm job, whether the commute is acceptable.

- The time it will take to get into town to gather supplies and basic necessities.

- The distance from farmers' markets, restaurants, food wholesalers, or slaughterhouses where you might sell your products.

- Noise, water, and light pollution—are you located near an operation that would create any of these?

- The time it will take for emergency responders (including veterinarians) to reach your farm.

- The proximity to car and tractor repair shops.

- Can you get mobile phone coverage? Mobile service is important as a safety tool.

- The location of schools if you have, or plan to have, children.

- How close are your neighbors? Too close or not close enough?

- Is the location feasible to sell products directly from your farm or a farm stand?

How Much Land Do You Need?

If you are looking to make approximately $50,000 a year from a farming venture, or what we'll call, "a living," there are some general rules of thumb for the amount of land it will take. Now, we don't know a whole lot of farmers who make that much money. So, if you are only looking for a second income, you can cut these numbers in half. A farm income of $25,000 (which in this book I call partial income) is fairly typical and we know many farmers operating at this level. As diversification is the route that I and most small farmers advocate, you might just be looking for a sideline or partial income. Or perhaps you just want to raise animals or crops to feed your own family as sustenance.

There are, of course, many variables that go into the acreage calculation, including the market for products where you live, your climate, the fertility and openness of your land, your own cost controls, and how you rotate your animals or crops. The following chart is a general rule of thumb:

- *CATTLE:* 200 acres for a living, 35 to 40 acres for partial income, or 5 acres for sustenance.

- *HAY:* 500 acres for a living. This is highly regional and this would apply to my area in Virginia. The market for hay around here is about $6 per square bale. Down in Texas in drought conditions, it's $8–10 per bale and it takes a lot more acreage to produce the same amount. A partial income can be had on 25 acres, but it's not cost-effective to maintain all the hay equipment for anything less. But you can grow grain for sustenance on a small scale.

- *FLOWERS:* 3 acres for a living, which would include some high tunnel or greenhouse structures; 1 acre for a partial income of $15–30,000 a year; or a lovely, smaller cutting garden for your own pleasure and to have a few bouquets to sell at the market alongside your other products.

- *VEGETABLES:* 4–5 acres for a living or anything less for a partial living or your own sustenance. Plan on $15–20,000 per acre, although some high-intensive growers in the proper climate can get $50,000 or more from 1 acre.

- *ORCHARD:* 20 acres for a living if you're growing highly desirable fruit, such as peaches. Anything less for partial income or sustenance.

- *NURSERY:* You might only need 1–2 acres to make a living if you have them covered in greenhouses and you're producing plants all year round.

- *HYDROPONIC GREENHOUSE:* 1 acre, depending on your crops

- *SUSTENANCE* (vegetarian): 1 acre

- *SUSTENANCE* (with animals): 3–5 acres

- *BEEKEEPING:* Can be done on a rooftop. Expect 30–50 pounds of honey on productive, healthy hives. That's $300–500 per year, per hive.

Taxes, Zoning, and Easements

One important thing to find out when purchasing a farm is whether the previous owner had an agricultural property tax exemption. Most counties offer farmers lower property taxes to encourage farming and discourage overdevelopment. As long as farmers are actively pursuing an agricultural business, they qualify for lower property taxes. To get the tax benefits, there are minimum requirements, such as keeping one cow, five sheep, or 25 chickens. Each county has different requirements, but the idea is that you have to have an active farm business to get the exemption.

Of course, many folks stretch the definition of *active farm business*. It's fairly easy to keep the minimum number of animals, sell one or two a year, and keep your exemption. Once you get an exemption, you never want to let it lapse. And if you're buying a farm that has an exemption, you should be prepared to immediately resume active farming. Otherwise, if you buy a working farm and don't farm on it, you will lose the exemption and be on the hook for higher property taxes—dating back to as many as five years, in some cases. You'd owe back taxes on the land for the time before you even lived there.

With the current domestic boom in natural gas and oil production, it is extremely important to ascertain the property's mineral and petroleum rights before you purchase any land. The title process will help you discover which rights will be included with your land. Water rights and surface rights are most necessary for farming. But it's the petroleum and gas rights that can create the biggest headache for you. If you do not own them, then an oil or gas company that purchases them has every right to come onto your property, build roads, drill wells, and erect fencing around those wells. Or they can drill horizontally under your property and endanger your water supply. On the plus side, if you own the rights, you can sell them, direct the exploration, and generate income that may pay for your entire farming operation, without sacrificing your well-being.

You'll also want to find out whether your property is zoned for agricultural use. This is only a question if you are living within a municipality or on the edge of a suburban area. Generally, in a completely rural area everything is zoned for agricultural use.

And lastly, you'll want to be aware of any easements that cross your property. An easement is a right to cross or occupy someone else's land. The electric and phone companies have easements under their power lines that allow them to maintain them. If those lines run right over your garden, you'll want to make sure that you keep them clear, to avoid the power company's coming in and spraying herbicides to keep vegetation from interfering with the lines. If you are trying to obtain organic certification, you'll need to work with your electric company on an agreement for "no spray" maintenance of the area around its lines. Environmental easements keep land from development and generally allow farming on property, as long as it's done sustainably. But you'd want to find out the requirements and make sure lots of bureaucratic steps aren't involved in working with an environmental easement on your property.

Insurance

Homeowner's insurance should cover everything you need on a homestead or hobby farm. But once you decide to run a farm as a business, you'll want to obtain a farm rider to your homeowner's policy. Generally, homeowner's policies only cover the house and immediate area around the house. A farm rider will cover all the other areas and needs of your farm, including outbuildings and barns. It offers you extra liability, usually up to 1 to 2 million dollars. You'll need this if you are selling products that people will eat and could become sick from; or if you plan to

have people coming out to the farm regularly, where they might injure themselves; or if you have animals that can escape your property and cause accidents or damage to others.

Urban Farms or Community and School Gardens

You should think about the discussions on finding and starting a farm even if you're exploring the idea of creating an urban farm, or community or school garden. Typically, you would not be buying the property where you are starting an urban farm or community garden, and certainly not a school garden. But you will need to make arrangements with the owner, and in many cases, a homeowners' association to use the land you want to farm. This, of course, limits your freedom, but it can also bring well-needed funding to get started. Many homeowners' and neighborhood associations and school boards welcome community gardens as a way to fight urban blight, beautify empty lots, educate, and offer local food to areas that might be underserved by markets. So, you may find that they can dedicate a portion of their budget to help you maintain the garden. They can also be a welcome source of volunteer labor.

Obtaining a good source of water can be challenging, however. If the lot you would like to farm on has no water hook-ups, it may be difficult to get service without having an actual dwelling there. Most likely, you would not have a well and would therefore need to use the city water service. One option is to befriend abutting neighbors and pay them a monthly fee to hook into their water. You can evaluate their monthly water bills to find out how much they are paying on average and then you can agree to pay whatever the increased cost is once you begin watering your garden.

Soil is another issue that needs attention before you get started, and not only because there may be very little topsoil on an abandoned lot. The lot could very well be contaminated. You should do extensive testing on your soil by an environmental service company or consultant. You can find these services easily in the phone book or online. The normal testing of soil done by the agricultural extension agency won't be enough. That testing is only for soil fertility and not any environmental dangers. Note that even if you find unhealthy levels of contaminants in the soil, you can still create a garden or farm. But you may have to invest in some remediation, excavation, and replacement of soil. This can be expensive, but not always. Sometimes all you would need is a truckload of topsoil to build raised beds. But rely on the professionals to advise you on what's best to create a safe environment to grow food.

Zoning is obviously an issue. This is one reason many urban farms and community gardens are nonprofit. Because they aren't operating as a business, they can get around some of the laws aimed at businesses. Zoning laws are being modernized all over the country to allow agriculture in cities and the suburbs. You can keep bees in New York and chickens in the suburbs of Washington, DC. You'll need to do your homework and decide what you can and can't do in your area.

- First Lady Michelle Obama's Let's Move! initiative promotes community and school gardens. She made news several years ago by starting a vegetable garden at the White House. You can learn about the school community garden aspect of her program on its website, www.letsmove.gov/gardening-guide.

- The USDA's People's Garden initiative started out as a challenge to employees of the USDA to create gardens at their facilities. The program has now grown

The subway runs above the garden at East New York Farm.

FINDING A FARM—BUYING, LEASING, AND INHERITING 67

to include 700 local and national organizations that have successfully established over 2,000 community and school gardens. You can find out more at the USDA website, www.usda.gov/wps/portal/usda/usdahome?navid=peoples_garden.

There are many different ways to live your farm dream. You do not have to own your land to be a farmer. And with so much in flux in the farm world, there are many opportunities to live and work on land that don't require you to buy it. And there are many ways to eventually become an owner without putting yourself into debt. Farming is creative. So, be creative and keep an open mind when you're working toward your dream. Try not to be wedded to one idea of what your dream farm should be. If you haven't been farming for a while, you may not know what will truly make you happy.

THE HEALTHY FARMER

The term *farm safety* immediately brings to mind images of working with large animals, operating a tractor or other heavy machinery, and using a chainsaw. The health and safety of our body benefit greatly from preventative measures. Living and working on a farm requires a level of physical toughness, but they also call for a truly mindful approach that limits the chances of bodily harm. It is important to remember that just because a task is routine and fairly simple doesn't mean it should be approached without care.

Louise Langsner at Country Workshops in North Carolina . . . taught me that how you conduct yourself as a human being really affects the things you create. Every day she did tai chi in the garden and took breaks to go swimming in the pond, and I swear you could taste her peace of mind in the tomatoes.

– RACHEL WILLIAMSON
Fairweather Farms, Afton, Virginia

Let's face it: Women and men are different when it comes to physical labor and safety. Women don't like to admit this, for the most part, but virtually every woman farmer I spoke to is very clear that there are general differences between the body of a man and a woman. Erica Hellen of Free Union Grass Farm says, "I don't think there's anything wrong with admitting that I'm not built the same way [my husband] is. He's got an extra one hundred pounds on me. Just the way our bodies work is different."

Besides the extra weight and leverage that weight brings, men tend to have a stronger upper body. All this really means is that you need to find different ways to approach certain chores. You want to know your limits and to adjust your posture and work habits to keep your body healthy and safe. In the chapter on tools, there are specific examples of tools that fit women well and that offer ways to compensate for a woman's upper-body strength and size.

Staying healthy on a farm requires more than just safety gear. There's a lot to learn about taking care of your body to avoid injuries; educating yourself about harmful plants, animals, and insects; planning ahead for common emergencies; and keeping children and other people from injuring themselves due to carelessness. Just as important as physical safety is your own mental, emotional, and spiritual well-being.

My mind thinks I'm still in my midtwenties, but my body is quick to remind me that it just isn't so. As I get older, all of the bending, lifting, and reaching takes a heavier toll than it used to. My knees make crunching noises and my neck and back are quick to point out when I've overdone it. Wearing sturdy and supportive shoes has become paramount to help avoid strained arches and painful calluses. Torn rotator cuffs, torn flexor tendons of the hands, hernias—these are some of the injuries women farmer friends have suffered. These injuries are incurred mainly from overusing particular muscles. We're strong and we push ourselves hard. We can't always avoid injury, but we can try.

After recovering from surgery for multiple hernias, which is a common farm injury due to all the strained lifting involved, Susan Parks (Broadhead Mountain Farm, Keswick, Virginia) says, "Pain is a great teacher. I am more aware of how I move and I try not to hold my body in one position for too long. Instead of bending over to pick beans, I get on my knees—I use knee pads

and shin guards. I try not to lift more than 10 pounds when I am bending over, and if I plant 16 rows of potatoes in the morning, I will take an Epsom salt bath that evening. I think farming is a little like motherhood. Nothing can really prepare you for the shock of how hard and constant and demanding it is. You can take classes, but when the responsibility is all your own it is, at times, overwhelming and scary."

To prevent muscle pulls and joint stress that could cost me days of recovery time, I do a series of simple stretches each morning. The 20 minutes I spend doing yoga is the preventative tool I use to keep from injuring myself doing my daily farmwork. It helps me stay just that little bit more flexible, and has made me aware of my muscles and bones so that I recognize when I'm pushing too hard and can back off before I get hurt. There have been times, mostly at the height of the growing season, when I've abandoned my yoga, telling myself, "It's only cool in the garden for a few hours—I need this extra time to get all these flowers picked and prepare deliveries." I always regret it. Not just physically, but mentally as well.

Yoga helps keep me flexible and strong, but it also helps clear my mind—for those 20 minutes, the stress of the farm business is not allowed to intrude and my brain gets rejuvenated for the day ahead. The measured breathing also helps prepare me for the rhythm of farmwork. Mindfulness and rhythm are important tools in keeping yourself healthy while doing the fairly monotonous work of cultivating, planting, weeding, and harvesting. It's very easy to get impatient or tired and then begin to hunch over or to lift with your back instead of your legs. Yoga helps me approach farmwork with a truly healthy and attentive attitude.

Yoga may not be your cup of tea. My sister takes a Pilates class that has strengthened her core muscles, important for protecting her body when lifting and bending, and another farmer friend takes the time to swim two or three times a week.

To maintain the energy and enthusiasm with which we entered our farming enterprise, we need to be mindful of our approach, completing our daily tasks with care, knowing when to push on and when it is time to take a break. It's true that often it seems there is no time to take a break. There is simply so much to get done. The day—even in midsummer with 16 hours of light—is short and the list of tasks is long. Getting crops in at the right time is crucial for their success, and succession planting is critical for a long and prolific harvest season. All the chores that go into accomplishing these goals are time and labor intensive.

Often this stress "that keeps you running from can until can't" can be more debilitating than many of our aches and pains.

A farm is a manipulative creature. There is no such thing as finished. Work comes in a stream and has no end. There are only the things that must be done now and things that can be done later. The threat the farm has got on you, the one that keeps you running from can until can't, is this: do it now, or some living thing will wilt or suffer or die. It's blackmail, really.

–KRISTIN KIMBALL,
The Dirty Life: On Farming, Food, and Love

A few strategies to help tame that unruly beast:

- Identify which aspects of your farm are most important to you. What are you spending time or money on, which doesn't really matter or work well for you? This can be as simple as not growing squash anymore because of the inordinate amount of time spent squooshing squash bugs and hunting for their eggs. I don't eat that much squash, anyway.

- Put systems in place that help you save time and labor. Wide, mulched pathways give me easy access to my plants and help suppress the weeds. Weeding always ends up near the end of my priority list.

- The idea of so much to get done is stressful in itself. One way to help get this crazy amount of stuff to do under control is to make a list or chart to prioritize what gets done and when. I include yoga and other exercise on this list, even though I sometimes bump them down a notch or two.

⇨ The weather is a major player in farm operations, so pay attention to the long-range forecast in your area. I check the National Weather Service online two or three times a week and jot down the temperature and precipitation forecast for the week. Then I can plan my days better and be ready to get a lot of work done early and late on hot days.

⇨ Let go of expectations of "the perfect farm." They stand in the way of our enjoyment of our real farm and promote dissatisfaction.

⇨ Take the time to relax with friends. The rejuvenating power of a conversation with a friend is one of the most valuable things you can do for yourself.

Posture, Position, and Breath

While you are working, it's as important to learn to position your body properly as it is to wear all the proper safety gear. To avoid injury, you must use your body correctly. And doing so will increase the strength you are able to put into tasks. While it's said that women have less upper-body strength, I'd argue that I have more than many 60-year-old male farmers that sit on their tractor all day, so I'm certainly not at a disadvantage. There's no inherent benefit in farming for big muscles. It's all in how you use them.

I try to use wheels every chance I get or think of clever ways to give a tool a dual purpose. Farming affords you a great many opportunities to be resourceful and creative. I really like that, and it keeps me from being bored.

–SUSAN PARKS

Broadhead Mountain Farm, Virginia

In a recent *New York Times* piece celebrating the 13-year old pitching sensation Mo'ne Davis, who can throw a 70-mile-an-hour fastball, Eric Anthamatten examined the origin of the derogatory term, "You throw like a girl."

According to political philosopher Iris Marion Young, the "girlie throw" results from a restricted use of lateral space that tends to come only from the localized part of the body that is doing the action—the hand and forearm—and rarely uses the whole arm, the whole body, or the extended space around the body that is necessary to execute the throw. Women "tend to concentrate our effort on those parts of the body most immediately connected to the task," she writes, and do not "bring to the task the power of the shoulder, which is necessary for its efficient performance." Think of the woman as she passes the pickle jar to the man to open. The inability to not open the jar has nothing to do with inherent strength, Young argues, but has to do with the utilization of the entire body for the task, something that is not rooted in anatomical or biological "limitations," but the whole social, political and aesthetic history of how females come to learn to "be" their bodies in space and time.[6]

Remember this story and the origin of the expression *throw like a girl*, and use it both as motivation and also because it points the way to avoiding injury. Use the strongest muscles

available to do each task. Everyone knows that you are supposed to bend your knees when you lift heavy objects or shovel
dirt. The reason is that you get leverage from the strongest
muscles in your body, your legs, instead of muscles in your back
that are prone to injury.

The same is true for digging with a hand trowel to plant
bulbs. Instead of primarily using your wrist to twist the trowel
in the ground, hold your wrist firm and turn with your shoulder
while keeping your elbow tucked close to your side. You still use
your wrist and your elbow, but you're now using a greater part
of your body—wrist, elbow, shoulder, and waist. You get more
power out of your effort while preventing injury.

Also, pay attention to the angle of the tool you're using.
Well-made tools have a proper angle for use to make them more
effective and prevent injury. Green Heron Tools in Pennsylvania,
for example, manufactures and sells tools that are specifically
made to fit a woman's body. The company's specially made
shovel has a wide handle that is easier to grip and a wide step
on the blade that makes it easier to push into the ground. When
you buy a tool, make sure to fit it to your body. There are various
lengths of tools and weights. Try them out and don't use tools
that are too large.

Just as important as using your muscles properly is to pay
attention to your breathing. It sounds obvious, but when you're
doing strenuous work, sometimes you can find yourself holding
your breath and straining. This not only will tire you out more
quickly, but it means that you aren't as focused and deliberate
with the work you're doing. Focusing on regular breathing and
rhythm while you work keeps you mindful of the task at hand and
helps prevent injuries—both muscle pulls and mental mistakes.

Keys to avoiding injury:

- Watch your posture—squat as much as possible instead of bending.

- Work from a kneeling position when you can, with knee pads, and straight back.

- Use the right tool for the job. For instance, a pick mattock is better than a shovel for digging a hole.

- Use tools at the proper angle—best at about a 70-degree angle to the ground so you are as upright as possible.

- Create a sustainable, even rhythm.

- Remember to breathe

COLD COMFORT

It's a miserable gray winter day in February. After 24 hours of steady rain, I know I have to go out and check on the pond drain. Grasses and other plant debris often clog the drain, causing the pond to overflow and erode our small dam. A breached dam would flood our fields, creating a wet and dangerous climate for our animals to live in for the cold and mud season. It will be getting dark in about an hour, so I can't put it off any longer. This is a chore usually reserved for Michael. But he's away on business and that leaves me as the only available pond plumber.

On go the rain pants, warm socks, winter boots, rain jacket, gloves—down to the pond I go. No sign of the donkeys, probably in the shed staying out of the rain. The llamas, reveling in the winter air under their downy fleeces, are out in the field and glance at me as I pass; my hood is pulled up over my head with my chin tucked tightly.

Well, great, the boat was left upturned and is full of water. I struggle to tip the water out, slipping in the mud by the bank of the pond. The curious llamas have wandered closer as they nibble at the sparse tufts of green grass. Once the boat is finally spilled of water, I search for the paddle. I pull the boat, a small dinghy, over to the dock, grab the paddle, and clamber in unsteadily.

I can see from the boat that the pond drain is indeed clogged, and can hear the sucking sound of the air in the pipe as the water struggles through it. I awkwardly paddle toward the drain, approaching it in a tightening circular path, as if stalking a skittish animal. As I get closer, I reach out to pull the strands of leaves and grass from the drain cover and the boat begins to tip; I sit back, my motion taking the small boat out of drain range.

I repeat my circular stalking and approach again, this time removing my gloves to get a better grip on the slippery plant material. The rain is running down my face and into my collar and the sucking noise from the drain is loud and hungry, and as I tip closer to pull away the debris, I feel my hand being tugged into the drain. I jerk away; the paddle slips out of my now numb hand and into the water. I grab for it, and within an instant I'm fully dunked in the freezing pond.

As I feel my boots fill up with cold, mucky pond water, I can't help but think that this was all planned by the farm gods—the tippy boat, the sucking drain, the freezing rain, and a couple of impassive llamas gazing at me from the field. Our

pond is not deep and I manage to keep my head above the water as I snag the rope attached to the boat and drag it, and myself, out of the pond.

I heave the boat out, turn it over, and leave it on the bank. Standing in the mud at the edge of the pond I remove my boots and pour out the water. Squeezing handfuls of water from my pockets, I realize the futility of trying to get less wet, put my boots back on, scrape my hair from my face, and search for the paddle. What luck: It's right there at the edge—things are finally starting to go my way.

On this day, the pond drain will not be unclogged. I realize that it's just too dangerous and it will have to wait for me to call in reinforcements or at least a lifeguard. I trudge, slow and soggy, up through the field. I wave a desultory hand at the llamas that are calmly watching me while offering no help or sympathy.

This story is meant as a cautionary tale. It is easy to read through it and list all of the mistakes I made. For one, it's never wise to set out on the water, even in a small pond, when you are the only person around. A breeched dam can be fixed; being sucked into the muck at the bottom of a pond is disgusting and scary at best. If you're in a situation where there's no option but going out on the pond, there are steps that you can take to make it safer: wear a life vest, call a friend to come over and watch, or at least remove heavy clothing and boots. Even if it's cold, it's a better alternative than struggling to stay afloat in them if you should unexpectedly enter the water.

BASIC FARM GEAR

WORK GLOVES Everyone has their own preference. The keys to choosing work gloves are comfort, utility and durability. A heavy pair of leather gloves offer durability, but aren't easy to use when dealing with small tasks. A thin pair of cloth gloves offer freedom of movement, but aren't very durable and certainly don't offer much in protection. Gloves are a tool, and like all tools, you should match the glove to the work you are doing.

The most recommended gloves by other women farmers are those with cloth tops and rubber on the palm. They wear out quickly, but are the most versatile and allow you to grip small objects easily.

My favorite pair—these leather gloves with flexible material at the knuckles are expensive, but very comfortable for long hours of work.

SHOES/BOOTS/FOOTWEAR

Comfort is paramount for the farm footwear you choose. Foot pain due to bunions and falling arches are some of the most common ailments for women farmers. Protection is also important. Heavy boots are required for logging and wood splitting. On the other hand, I know women farmers who prefer bare feet in the garden. The gardens they work in are well-tended, however, and have little in the way of rocks and debris. As with gloves, match the footwear to the task at hand.

My favorite farm shoes, made by Keen. They have strong leather and rubber uppers for protection, extra arch supports and are slip-ons for convenience and comfort.

These leather ankle boots work well around animals. They offer enough protection from hooves and muck, with shallow treads so you don't track manure very far.

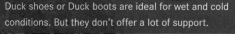
Duck shoes or Duck boots are ideal for wet and cold conditions. But they don't offer a lot of support.

Every farmer needs some rubber boots or insulated rubber boots.

Heavy leather boots are needed for cutting trees and splitting wood. Steel-toed boots are preferable.

Rubber shoes or Crocs are popular farm shoes, but offer little in the way of support.

HAT A wide-brimmed hat is a must when working outside. Make sure the brim is wide enough to offer shade to your ears, nose and neck.

LONG-SLEEVED SHIRT White is the ideal color to keep you cool while reflecting the sun. And the more skin you can protect from bugs, the more comfortable you will be.

KNEEPADS Kneel as much as possible when working with low-growing plants to help prevent back injuries. Kneepads and shin pads provide lasting comfort for this kind of ground work.

BUG SPRAY & SUNSCREEN You're already going to accumulate cuts and scrapes along the way, so protect your skin from sunburn and bug bites.

GETTING TO KNOW YOUR FARM
FROM HOUSE TO FIELD

CHAPTER   FIVE

Part of getting to know your farm is getting to know your community. You aren't just making sure you have a good base to buy your goods at market, but you're also building a nice support group for your new venture. Some of the easiest ways to find like-minded people are to buy local products, shop (or even work) at locally owned businesses, and browse, buy, and eventually sell at the local farmers' markets. Attending workshops and classes offered by local agencies and farms is another way to get to know those who share your interests, as well as increase your own knowledge and awareness.

This said, if you've decided to make the leap and move out to the country to have more space for your farm, life in a rural setting can be isolating. In most cases you are at least 20 to 30 minutes from a grocery store, post office, bank, and pharmacy. Unless you make a trip to town—or take advantage of a workshop or class, as mentioned above—you might not see people for days. This might be a good fit with your personality—it is for mine, as I enjoy spending time on my own. As I've grown older I've gone from being somewhat extroverted to mostly introverted. I am never bored and rarely pine for the company of others. As I don't need to be around other people so much, life on an isolated patch of land is not a hardship for me. If that isn't your personality, though, you may want to consider staying closer to town; sacrificing some land in exchange for your personal well-being might be well worth it.

In either case, knowing and talking to neighbors who are living and working on a small farm is invaluable. I've learned so much about farming and living on a farm from those already living that lifestyle (such as where to buy things I need, what kind of tool/machine is best for certain jobs, what products, manufacturers to avoid, procedures for certain tasks and emergencies), as well as gained through their introductions to others with interests similar to mine, and I don't think I could live on a farm without knowing that I have their support. I've tried to give as much as I receive, and I help my new neighbors with their gardening and country skills questions, as well as listen to their concerns and successes. So, the most important thing to get to know about your farm and homestead is your neighbors.

This chapter focuses on the basic things you'll need to know about most farmhouses, from the wells and septic systems, to the maintenance of woodstoves and pipes, to preparing for winter and handling power outages. But all the manuals in the world aren't as valuable as the knowledge living right next door (or nearby, anyway). Your neighbors are the resource to which you will inevitably turn. So, get to know them. You don't have to be best friends. But you're out in the country together and just knowing that you have a safety valve nearby provides important peace of mind when you're far away from immediate emergency services. Community is your primary safety valve when you're in the country.

And when something goes wrong out in the country, especially with your house, it will probably be a neighbor who's first on the scene to help you out. So, while you learn a few things about how your farmhouse works and how to keep it maintained, remember that you can always call for help if you've made that connection. They've probably done all the things described in this chapter and will hopefully be happy to share that knowledge with you or at least provide some moral support and advice as you undertake the care of a farmhouse yourself.

One of my inspirations for writing this book was *Dare to Repair*, by Julie Sussman and Stephanie Glakas-Tenet. Their book is a guide for women to fixing just about anything around the home, from repairing leaky faucets to lighting a gas furnace. You should buy that book and keep it as a reference. But even if you aren't comfortable fixing everything right away, there's an easy way to begin learning how. Whenever you have repair or service people come out to your home, ask them a lot of questions, watch what they do, and take notes. Treat the service call as a tutorial. You are paying good money for it, after all. This is how I learned to brush out the stovepipe, adjust the pressure on our water tank, and change the filter on the heat pump. Servicemen are suckers for women who are interested in what they do. Ask questions and eventually you won't have to call them out again (unless they are cute, of course), because you'll feel confident doing it yourself.

Wells

If you're living in the country, most likely you are getting your water from a well. Folks in the city rely on water to be pumped from a treatment facility into their home. In the country, most people get their water straight from the ground. This is possible with a well. You're certainly familiar with the old-time wells you've probably only seen in a movie. They were about 2 feet wide, dug into the ground and lined with rock. There was a well house on top and you lowered a bucket down to where the water line was and brought it up a bucket at a time. Kittens were drowned down there, or so you heard, and sometimes bodies were dumped there.

Now, we have high-tech drillers that come out with their large truck and bore a well as deep as 300 to 400 feet. It all depends on where you live and how deep your water table is. The deeper the well, the less chance it will ever be polluted. To get the water from the well into your house, you will have an electric well pump. This pump, placed in the bottom of the well, brings water up to your house to a pressure tank. The pressure tank takes the water pumped up from the well and pushes it into your hot water heater and through all your pipes in your house to your faucets. This is how you have pressurized water when you turn your faucet on.

To keep your water flowing and clean, there are a couple of maintenance chores you should undertake every year to every other year: (1) check and maintain the pressure in your pressure tank, (2) test your well water, and (3) disinfect the well if the test shows signs of bacteria.

Testing the water for bacteria is part of the legal requirements when you buy a home, so you will know right away if there's an issue. When a well is tested for bacteria, they only test for coliform bacteria. This is common bacteria; if present in low levels, is not

An example of a drilled well.

generally dangerous to humans. But its presence indicates that other, more dangerous bacteria might be present. Your water testing facility will let you know whether you have levels that are concerning and whether your well should be "shocked," which is another term for disinfected. And again, if you are uncomfortable doing this the first time, call a professional to show you and then you can use these directions the next time you do it.

How to Test the Pressure in Your Pressure Tank

◆ If you are at all uncomfortable performing this from these instructions, call a professional to come out and show you how to do it. A service call should cost less than $100 and once you've seen how easy it is, you'll feel confident in doing it yourself.

◆ First, locate the tank. It should be right near your hot water heater. It's usually a blue or gray tank about 3 feet tall.

◆ Turn off the water and drain the tank. To do this, you need to find the circuit breaker for the tank and shut it off. Then open several faucets in your house and wait for the water to stop coming out. This may take a couple of minutes. Shutting off the water is a good thing to know how to do, as you need to do this anytime you are changing a faucet, working on a toilet, and so on.

◆ There will be a pressure outlet on the top of the tank that looks like the one you'd find on a tire. Test the pressure just as you'd test a tire with a tire pressure gauge.

◆ Check the proper pressure that your tank should have, by reading the label on the tank.

◆ Use an air compressor (the same kind you'd use on your car) to bring the pressure up to the desired level.

◆ Turn the power back on and turn off your faucets.

How to Disinfect Your Well

- Mix 2 quarts of bleach (must contain sodium hypochlorite and no perfumes or scents) in 10 gallons of water; pour into the well.

- Connect a garden hose to a nearby faucet and spray down the inside of the well until you can smell bleach.

- Open each faucet (your sinks, showers, anything that runs water) and let the water run until a strong chlorine odor is detected, then turn the faucet off and go to the next one. Don't forget outdoor faucets and hydrants. Drain the water heater (by running hot water in a tub) and let it refill with chlorinated water once you get a strong smell of bleach from the hot water in the tub. If a strong odor is not detected at all outlets, add more chlorine to the well.

- Flush the toilets.

- Mix an additional 2 quarts of bleach in 10 gallons of water. Pour it into the well without pumping.

- Allow the chlorinated water to stand in the well and pipes for at least 8 hours (preferably 12 to 24 hours).

- Run water from an outdoor faucet through a hose and away from your house, making sure it runs away from desirable vegetation, until the chlorine odor is slight or not detected at each faucet. Then run indoor faucets until there is no chlorine odor. Minimize the amount of chlorinated water flowing into a septic tank. Bleach is bad for septic tanks as it disrupts the natural breakdown of the waste. So, minimize the amount that goes down your drains by running as much outside as possible.

- Some chlorine may persist in the system for 7 to 10 days. Water with a slight chlorine smell should be usable for most purposes. If the odor or taste is objectionable, simply let the water run until the chlorine dissipates.

Approximately two weeks after flushing the system, sample the water (according to laboratory instructions) and have it tested for biological contamination. Repeat the test in two to three months to be sure the system has not been recontaminated.

If water tests show that biological contamination has reappeared or persisted, try to locate and remove the source of bacteria. Human and animal wastes are common causes of bacterial contamination, so a nearby septic system or livestock pen could be the source.

If the follow-up water test shows no bacterial contamination, you should still test your water once a year. If there is a change in the taste or smell of your water, or if there are unexplained illnesses in the household, test the water as soon as you notice the change.[7]

Septic Systems

If you get your water from a well, you most certainly have a septic system to take care of the waste. Water and waste from your sinks and toilets leaves your house in a pipe and goes into a septic tank. It's usually located in an open area, such as your yard, which is free of trees. You can find your septic tank on the plat drawing of your property. You should have that if you've bought your house. If not, then ask your landlord or previous owner.

The waste flows into a big, concrete tank located underground and begins to break down naturally with bacteria. As the level rises, the top layer is turned to liquid. When this liquid reaches a certain level, it drains out of the tank through a pipe to another smaller tank called the distribution box. When that tank fills up, the waste that's now completely liquefied flows via gravity down long pipes that fan out across your yard or field and then is naturally absorbed in the ground through holes in those pipes.

It's really a simple and ingenious way to deal with human waste and incorporate it back into the earth. But old septic tanks that are not maintained are a big contributor to water contamination. If your well is continually showing contamination, your septic system should be one of the first things you check.

What do you do to keep your septic system working properly? The better question is what *shouldn't* you do? You should not have a sink disposal that puts food into your system. Food disturbs the balance of human waste bacteria in the system and can contribute to clogged pipes. You should also avoid putting bleach or any powerful chemical cleaners down any of your drains as it will kill the bacteria you need to break down your waste. Only buy cleaners approved for septic systems. You should also not buy triple-ply, baby butt–soft toilet paper. The thicker your toilet paper, the longer it takes to break down and the easier it will be for it to clog up one of the pipes. And once you locate your system, don't drive over it or plant trees or deep-rooted plants that might disturb the underground pipes or tank. Some people use a bacteria treatment once a month or so. You can buy it at the grocery store and you just pour it down your drain. It's supposed to help the good bacteria grow. We've never used it and never had a problem.

And while we've only had our tank dug up and pumped out once in the 12 years we've lived here, it's generally recommended that you have a septic technician pump out your tank once every three years. There are newer septic systems that have an electric pump that circulates the treated waste in a long pipe. These systems need to be serviced once a year and are dependent on electricity. You'll need a trained technician to service it and let you know what to do to keep it functioning properly.

Wood-Burning Stoves

I love my wood-burning stove. Is that wrong? I guess it's not my stove so much that I love, but the heat that it puts off. Our stove came with our house. It's what you would call a firebox, more than a stove. A firebox is really just a metal box that is lined with stone inside. You burn a fire in it and it heats up everything around it. A better stove is one that has a catalytic converter that recirculates the air. In this way it takes the fumes from the burning wood and then burns them again. It's more environmentally friendly, as you burn most of the soot before it escapes your stovepipe and it's more efficient because it generates more heat from less wood. Add a blower fan to it and you've got one heck of a nice stove. As you can imagine, the firebox that we have is much less expensive than the Cadillac, catalytic converter stove.

Not everyone needs a stove, for sure. But with temperatures reaching freezing in even the most southern parts of our country now, why not have one for just in case? If you don't have one now, go shopping for one and ask a lot of questions. Like everything else, your ultimate decision will probably come down to what you can afford. You should of course have a professional install it for you and that's usually incorporated into the price if you are buying from a local dealer.

Considerations when buying a wood-burning stove:

SPACE if you only need to heat your living room or give yourself supplemental heat, then you don't need an expensive stove. But if you plan to heat your entire house, then go with the best you can afford. Ignore the Btu numbers the manufacturers provide. They are misleading.

THICKNESS OF THE METAL You can get a welded steel or cast iron stove. There's not much difference in heat. It's the thickness of the metal that matters.

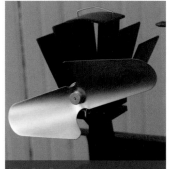
The Eco Fan runs on the heat from the stove and circulated the heat throughout your room.

Our wood-burning stove. An iron kettle full of water helps generate some humidity.

SOAPSTONE LINER this option traps heat and keeps it radiating longer, but it also take longer to heat up. So six of one, half a dozen of another.

CATALYTIC COMBUSTION newer stoves offer this technology which reignites the fumes from the initial fire. It's more efficient, environmentally friendly and cuts down on creosote buildup.

ORIENTATION AND SIZE some stoves load from the front and some from the side. Loading from the side tends to be easier. Also, you might be tempted to buy the longest one available thinking that longer wood burns better. Remember that you have to cut, chop and haul the wood. So shorter lengths are recommended for most women woodchoppers.

Stack your wood off the ground. We use pallets. You can stack wood between trees or drive posts in the ground for support. You can also chimney stack the end like you see here and then stack wood up against it. The key is to stack it as tight as possible. Think of it as a jigsaw puzzle.

ASH PAN a stove that has an ash pan that easily slides out to be cleaned is a recommended feature.

BLOWER VS. ECO FAN Blowers add a big expense to stoves. Buy an Ecofan (pictured) instead. It sits on the back of your stove and the heat drives the fan to blow air into the room. It's aesthetically pleasing as well.

AESTHETICS the cast iron stoves are more decorative, while the welded steel stoves are more industrial looking. If they are new, they both meet all the EPA and efficiency standards, so it's all a matter of taste.

If you have a stove already, you should first have a professional chimney installer or sweep come out and inspect it for you. You never know if it was the previous owner's second cousin, Billy Burnout, who installed it. And that's just not something you want to guess at. Once you have your stove installed and inspected by a professional, it's time to get burning.

What type of wood you burn all depends on where you live. We're lucky to have acres of oak and maple, which are some of the best "hard" woods. They burn efficiently and put off less creosote. Creosote is a tarlike substance that results from burning wood. Softer woods that are high in sap content, such as pine, put off more creosote than do hardwoods such as oak. But they all put off some and that is why your stovepipe becomes coated in a black substance. It's highly flammable and can catch fire in your stovepipe if you let it build up too much, so you need to check your pipe once a year and clean it out if there is any buildup. The drier your wood, the less creosote will be produced. It's best to burn hardwoods for this reason. But even if all you have is pine, just make sure you have seasoned your wood well so it's very dry and you will avoid much buildup.

Types of wood in order of quality for burning in a stove:

- Oak
- Elm
- Maple
- Hickory
- Ash
- Birch
- Cedar
- Pine
- Spruce

Again, depending on where you live, you may only have access to a certain kind of wood. In New Mexico, all we could get was cedar and pine. Any wood will burn; just make sure it's dried properly. Even wood that isn't completely dry will burn, but not as well and it will put off more creosote. But you burn what you've got and there will be times you're cursing at the fire for hissing and sparking and not putting off much heat. Plan ahead and cut firewood early in the spring to have it dried by winter, at the earliest. Wood that's been seasoned for over a year is best.

WINTERIZING

If you live in a cold climate, you'll want to prepare your house for the winter.

Here is a checklist to knock out while it's still just cool outside:

- Change the filters on your furnace or heat pump.

- Have your furnace serviced by a professional (don't forget to ask questions).

- Close the vents around your foundation.

- Drain and disconnect hoses from outdoor faucets.

- If you have outdoor faucets that are prone to freezing and they have cutoff valves, cut off the water to them and put a thermal cover over each faucet.

- Reverse your ceiling fans. Most have a switch that allows you to run the blades backward. This helps push around the room the hot air that gets trapped on your ceiling.

Cleaning out your Stove and Stove Pipe or Chimney

Cleaning out your stove and stove pipe or chimney is a job that you can do yourself. But if you aren't comfortable at first, hire a chimney sweep and follow him/her around and ask questions. The frequency of cleaning depends on how clean your stove burns. Keep the ashes in the actual stove from piling up too much or they will hamper the fire. Use a metal bucket (never use plastic) and scoop the ashes out when they are cool. Then set the bucket outside away from anything that might catch fire if any burning ashes were to alight in the wind. Only dump the ashes into your compost or elsewhere after at least a week of cooling. If you don't want to wait that long, mix them with water before discarding.

We clean our pipe out once a year before we burn our first fire of the season. I have some friends with very efficient stoves who only clean their pipe every two or three years. It's best to take a look down the pipe every six months or so during the burning season and check to see if there's creosote buildup (it will be a layer of black, flaky material). If there's a layer a quarter inch thick or more of creosote buildup, it's time to clean it.

PROCEDURE FOR CLEANING OUT YOUR STOVE AND STOVEPIPE

1.
After making sure your stove is cool and ready for cleaning, use a sturdy ladder to climb onto the roof.

2.
Assemble the wire chimney brush and flexible rods. Buy a brush that fits the diameter of your pipe.

3.
Loosen the pipe cap and remove.

7.
The brush also has a loop on the end that you can attach a rope to and drop down the pipe. This allows a person at the bottom to pull down while you are pushing. If you have angles in your pipe, this makes the job easier.

8.
Remove your pipe from the stove and remove the damper. Clean damper with wire brush. Newer, cleaner burning stoves do not have a damper to clean.

9.
Clean all the areas and openings that the pipe brush cannot get to by hand with a wire brush.

4.
Inspect the pipe. It's evident that this pipe has creosote buildup and needs to be cleaned.

5.
Use a wire brush to clean the pipe cap.

6.
Insert the brush into the pipe. Push and pull the brush through the length of the pipe to scrape off the creosote.

10.
Scoop out the ashes and other debris from the stove into a metal bucket.

11.
This stove pipe is now free of creosote and ready for a fire. Reassemble the pipe and pipe cap.

12.
To keep birds from getting trapped in your pipe, wrap chicken wire around it. Use only the chicken wire with wide holes or creosote will build up and choke off your pipe.

Generators

We've never used a generator. Sure, there have been times when we could have, but we didn't and we're still alive to tell about it. Many people use generators as security blankets. The thought of losing power is terrifying to them. But, really, although the power does go out a few times a year out in the country, it's never for too long, at least where we live. Of course, if you are living where the temperatures can be in the negative double digits, a generator would be a very good idea.

When Hurricane Isabel came through, Michael was out of town. The storm was scary and trees blew down and the power went out for four days. I did not have a generator and I did not end up needing it. I had rain barrels that I used to fill the toilets and I even bathed using them. Yes, some food did go bad in the refrigerator, but I also ate much of the rest.

I now have a gas range, so even if the power is out, I can cook. With my wood-burning stove, I can keep warm and boil water to drink. We've used the wood-burning stove to cook on as well. I have oil lamps and electric lamps and flashlights. And there's so much food sitting in my pantry that I could last weeks without electricity.

That said, we bought a generator this year. We bought it just in case the power goes out during the growing season, when we might have two coolers full of flowers for market. A power outage right before a big wedding could be a disaster for my flower business. So, we bought a generator as insurance for my business.

You should make your own determination about whether or why you might need a generator. If you do decide you need one, the first thing you should do is talk to an electrician or other farmer who's knowledgeable about them.

Advice About Generators

➧ The size of generator you need depends on what you will run off it. Talk to an electrician and find out the proper size and power.

➧ Never run a generator inside the house or in a basement. That's like running your car in a closed garage.

➧ Most generators are not big enough to power your whole home. You can only run a few extension cords from the house to the generator.

➧ You will not be able to power your well pump or pressure tank unless you have a special plug installed by an electrician for this purpose. This is a key mistake people make. Because you can only plug things into a generator, you need a plug. Only an electrician can run an outlet that serves such things as your water pump and heat pump.

➧ Try to buy the quietest model of generator you can find. Honda makes very quiet generators. Nothing is more annoying to you and your neighbors than the constant hum of a generator for hours on end.

➧ Always drain the gas from the generator after use. Or run your generator for about 20 minutes every month. Otherwise, the gas will go bad in the lines and you will not be able to start it when you most need it.

TOOLS AND TECHNIQUES

CHAPTER SIX

There's a tool for every task on a farm. The dirty little secret behind the farming mystique is that you don't need decades of farmer wisdom or exceptional physical strength to do all the basic chores on a farm. "Farming is so much work!" You will hear this exaltation incessantly from nonfarmers when you tell them about your plans to take up farming in whatever way you choose—whether it's a weekend homestead, a summer internship, or your very first purchase of a piece of land.

The truth of the matter is that with a few select tools, many of them hand tools, and a basic knowledge of the proper technique for using them, you can manage a farm with the best of them. The first thing to remember is that you don't need to do everything yourself. It's okay to delegate. A farmer girlfriend of mine operates an herb business on her partner's mushroom farm. One thing she has a real aversion to is chopping wood. So, she happily handles cooking for the two of them as long as her partner keeps that firewood cut, dried, stacked, and burning in their stove all winter long. If there's a chore or farm task you don't feel comfortable taking on, find someone who is willing to do it and trade whatever you can bring to the table, so to speak.

Another secret that I've learned from my husband is that most of the knowledge that men know about engines and tools comes straight out of the owner's manual! Men are not born with some kind of advanced knowledge of mechanics. When something breaks, they crack open the owner's manual and try to figure out what's wrong. Then after four trips to the hardware or auto parts store and after much trial and error, they finally get

it right. And then they go drink beer with their friends and talk all about it as if they just caught a 6-pound trout. So, read your owner's manuals. They always have very clear directions about everything you need to do to maintain your tool. You don't have to follow the maintenance requirements exactly. I don't know anyone who does.

Even if you don't take on all the wood chopping, fence building, bed laying, tractor driving, and engine maintenance, it's a good idea to know how, just in case. Regardless of where you live, there may be times when you need to fend for yourself. One errant winter system can bring unprepared electrical grids to a halt or drop a tree across your driveway. You'll want to at least learn some basic skills that will help you keep yourself warm, safe, and fed during any situation.

Here are some basic country skills that everyone who wants to be self-sufficient should know:

- Operating a chainsaw and chopping firewood
- Basic small engine maintenance
- Fixing flat tires
- Fencing
- Proper use and care of hand tools, including sharpening
- Tractors—full-size and walk-behind

I bet you thought gardening knowledge or animal husbandry might be included in this list. Those are optional skills for whatever you choose to do on a farm. But before you produce or care for anything on your farm, you should know how to take care of yourself and your basic farm infrastructure. Once you're comfortable taking care of yourself, you can be confident in learning to care for any plant or animal. While a couple of cats might help keep you warm, the same can't be said for a horse or a bunch of kale.

In this section, I'll guide you through the basic skills and tools to keep your farm running. In later chapters, we can explore further skills and tools that you'll want to use in the garden, field, and barn.

Operating a Chainsaw and Splitting Firewood

Safety is the number one concern when operating a chainsaw and chopping firewood.

When using a chainsaw, you should wear:

- A hard hat, as limbs can fall off trees when you're cutting them down

- Heavy, preferably steel-toed boots. It's easy to hit your foot with the saw, so the heavier they are, the less chance the saw will cut through your boot. Also, it's common to drop logs or pieces of wood on your foot. Steel-toed boots can be hard to find in women's sizes. If you order boots online, order a men's version that is two sizes smaller than your women's size.

- Eye protection

- Ear protection

- Logging chaps. These are either leather, Teflon, or some other strong material that will stop a chainsaw from cutting into your leg.

- Gloves

- Clothes that are not so loose that they will snag on branches or get caught in the saw

Other considerations:

- Cut only trees that have a clear path to fall and are not hung up on other trees or growing at an odd angle.

- Cut trees that are convenient to get to with your truck or tractor so you don't have to carry wood by hand for long distances.

- Practice with small, thin trees. If they are thin enough, you may not even need to split them before burning in your fireplace.

- Follow the owner's manual for your saw to make sure the chain is tightened properly. A loose chain will not cut well and can come off the saw. Check chain tension often.

- Do not cut trees if it's windy or icy. Wind can make trees fall where you don't want them and ice makes dangerous footing.

- Never operate a chainsaw if you aren't feeling alert, and certainly not after drinking alcohol or taking drugs or cold medications.

- Leave your pets and children behind.

- Make sure to clear a path of escape before you begin cutting.

- Do not operate a chainsaw above shoulder height or while standing on a ladder.

- Make sure someone knows where you will be and what time you expect to be back.

- Bring a cell phone with you if you have coverage.

While this is the basic technique for felling, limbing, cutting and bucking a tree, there are many things that can go wrong along the way. The tree can be pushed by the wind and lean in a direction you weren't planning and pin your saw in the tree. Or once the tree is on the ground, you might get your saw pinched and not be able to get it out. Felling wedges can help in these instances. Hammer them into a cut to keep it open. Use a straight, long crowbar to lift a felled tree to free a saw blade. It's difficult to describe all possible outcomes. It will take time and experience. That's why you should go out with someone who has experience several times before you attempt to do this by yourself.

STEP-BY-STEP INSTRUCTIONS FOR OPERATING A CHAINSAW, FELLING, LIMBING AND BUCKING A TREE

1.
Choose a tree. The dead leaves on the top of this tree show it's a good candidate to be turned into firewood.

2.
After clearing any obstructions around the tree, make the initial cut parallel to the ground, perpendicular to where you want the tree to fall and about ⅓ of the way through the tree.

6.
When you feel the tree start to vibrate and it begins to fall, retreat back and away. Don't forget to yell, "Timber!"

7.
Limb the tree by cutting branches on the opposite side of the tree to where you are standing. Start at the bottom of the tree and work your way up. Be careful as the tree can be leaning on limbs that will snap back at you as you cut. It's best to cut the limbs that are free in air and then turn the tree with a timber jack or peavey to get to more.

3.
Make the second cut at about a 45 degree angle down to the first.

4.
Make sure the wedge is fully cut away from the tree.

5.
The felling cut is made parallel to the initial cut, but a couple of inches higher so that it forms a hinge as the tree falls. Make sure your saw is perpendicular to the direction you are felling.

8.
Score the trunk by cutting lengths along it. Cut as far through as you can before the tree begins to pinch your saw. Don't go all the way through or you'll hit the ground or pinch your saw.

9.
Use a timber jack to turn or lift the log to get to the underside that you haven't cut.

10.
Finish the cuts (buck it up).

One of the issues with pull-start power equipment is that the machines can be difficult to start. Chainsaws especially take a good amount of shoulder strength to get going. Stihl offers various chainsaws and power equipment with Easy2Start™ pulls that even a torn rotator cuff can handle. Ask for them at your local Stihl dealer.

Shock your male friends by substituting canola oil for the bar-and-chain oil in your chainsaw. It's perfectly safe, a lot more environmentally friendly, and cheaper, and smells better when it burns. Tell them Martha Stewart taught you that trick!

Start a chainsaw on the ground with the chain free of any obstructions. Follow the directions from your instruction manual about operating the choke and primer. Insert your foot into the handle opening to hold the saw down. Place your left hand on the upper handle to keep it steady while you pull the cord with your right hand.

Splitting Firewood

The best tool to use to split firewood is a mechanical splitter. These are usually gas-powered or hooked up to a tractor. You can mechanically split even the toughest wood in seconds. They are expensive, however, and not always practical for a small farm. But if you have a friend who has a mechanical splitter, it's never a bad idea to borrow or trade for a day. You can cut several cords of wood in just a few hours.

If you are splitting by hand, you'll want to use a maul. A maul is a cross between an ax and a sledgehammer. The head is flat like a sledge on one side and comes to a blunt point like an ax on the other. A maul has the heft to really bust wood apart. An ax is really for cutting trees down the old-fashioned way, unless you're an experienced user and have an expensive ax.

Splitting wood by hand is good exercise. And it doesn't really take all that much strength once you learn to swing properly. It's best to have a chopping block to raise the log off the ground. Find a big, uncut log that is below knee height. If you don't have a piece big enough, ask another farmer if they have a nice big flat piece of wood that can act as a chopping block. Mauls come in different weights. The lightest one is four pounds, so start there if you aren't comfortable with a heavier one. The heavier the maul, the easier it will split the wood, though. Take some practice swings only using the chopping block to get a feel for it. Then start with short pieces of wood, which split easier, and work your way up to big ones. One key is to make sure you follow through with your swing. If you are timid and pull up without following through, you can miss your target and end up with the head of the maul coming back toward your shins. Let the weight of the maul do most of the work, but always follow through.

SPLITTING FIREWOOD
BY HAND

(Left to Right): 4-lbs. maul, 6-lbs. maul & ax.

1.
Practice on the chopping block first. At the end of your follow through, your hands should be even with the head of the maul.

4.
At the top of the swing (does not have to be all the way over your head), slide your dominant hand down to meet your other hand as you swing.

5.
Follow all the way through the swing as if your intention was to end at the top of the chopping block.

2.
Stand with your feet squarely to the target. Put your dominant hand under the head of the maul and the other at the base of the handle.

3.
Raise maul over your dominant shoulder.

6.
For big and knotty wood, you may need to use a splitting wedge and sledgehammer. Just hammer the wedge into the log and then sledgehammer it until the log splits. Once it's split in two, you can usually finish splitting the rest with a maul.

When splitting firewood, you should wear:

- Eye protection
- Heavy boots
- Gloves are optional. Some people prefer the tighter grip you can keep on an ax or maul when using bare hands.

A felling wedge can be knocked into the initial cut with the back of an ax to keep the cut open while you continue to saw.

Work a wood splitter with two people. One person lifts the logs onto the splitter. The other operates the splitter. Always keep your hands on top of the log and never on the ends.

Basic Small Engine Operation and Maintenance

Learning to do basic engine maintenance is an absolute necessity if you want to farm as a career. It's too expensive to have professionals do the basic work for you. And mechanical things on a farm break all the time. The more you learn about maintenance and repair, the better. Try to find classes at your local community college or at the least find a mentor who can help you along until you're comfortable.

As was mentioned early, the owner's manual is your friend! Read it for each engine/tool you have and find out the suggested frequency of maintenance and also the specific parts that you will need. When you go to the auto/farm store to buy oil, oil filters, pipe fittings, and so on, it's best to take the old ones with you, as there always seems to be some question as to whether you're buying the right replacement part. You should definitely write down the make and model of whatever machine you are working on, but taking the actual parts with you in a plastic bag is also advised to save yourself multiple trips.

All engines are basically the same, whether you're working on your lawnmower or your truck. The yearly maintenance requirements are the same, too.

First, identify whether the engine you are working on is a two-stroke or four-stroke engine. Not many people know the exact technical difference between the two (I don't), but you should know how to identify them. Two-stroke engines are smaller and require you to mix oil in with your gas. A weed whacker and a chainsaw will have a two-stroke engine. A lawnmower, riding lawnmower, and tractor will likely have a four-stroke engine. You put gas and oil in different tanks. So, that's how you tell the difference: if you mix your gas and oil, it's a two-stroke. If you fill the gas and oil in separate places, it's a four-stroke.

STORING YOUR GAS

Use multiple gas cans, so you don't have to lift the large, heavy ones. Several 1- or 2-gallon cans accomplish the same thing as a 5-gallon. Label them with permanent marker as to whether they are *regular unleaded* gas, which goes in your four-stroke engines, such as your car or lawnmower, or a *mix* of unleaded gas and oil, which goes in your two-stroke engine, such as your chainsaw. Use only a 1-gallon container for your mix, as it's easy to buy individual small containers of oil that are mixed with 1 gallon of gas.

BASIC MAINTENANCE OF
LAWNMOWERS

1.
Drain the oil. On small engines, you can just turn the machine over. Make sure to tape the gas cap closed. And put a tarp underneath to catch any oil that will drip.

2.
On larger machines, you will find the oil drain plug and unscrew it to drain the oil. Make sure you have a large enough container to catch the oil draining out..

5.
Cut away anything that might have become tangled in your lawn deck and clean out any debris sticking to the side.

6.
Install the new blade.

10.
Use silicon or other grease spray to lubricate the wheels and other areas that might squeak.

11.
Refill the oil until it shows full on the dip stick. Do not overfill.

3.
Larger engines have an oil filter. It's a short cylinder on the side of the engine. You will need an oil filter wrench to unscrew it and some oil will drain out of it. Once all the oil is drained, replace with a new filter. Screw it on by hand and not too tight.

4.
For lawnmowers and tractors, you'll want to change or sharpen the blade every few years. Unscrew the blade with a wrench.

7.
Find the air filter. Unscrew the lid.

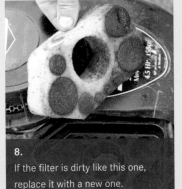

8.
If the filter is dirty like this one, replace it with a new one.

9.
Find and replace the spark plug. Unscrew with a wrench or use a socket wrench. Most socket wrench sets have a socket designed to fit exactly on a spark plug.

12.
Larger machines like lawn tractors have grease fittings. It's the small nub you see here. Your manual will show you all the spots where they are located on your machine.

13.
Use a grease gun and squirt only once into each fitting.

Basics of Starting Any Engine

→ Check the oil and gas levels and fill them up. Forgetting to check the oil and running out is a fatal error for an engine. Always check, as you never know when you might have sprung a leak.

→ Pull the choke all the way out.

→ If there is a primer button, press it five times or however many times are suggested on the engine. The primer pushes gas into the fuel line.

→ Make sure it's in start mode.

→ Pull the start cord or turn the key.

→ If pulling the cord, keep pulling until the engine sounds as if it turns over and dies. Then turn down or push in the choke halfway. Pull again until the engine starts. It will be sputtering a bit. Turn down the choke all the way and hopefully the engine will rev up. You might need to play with the choke a bit. If you turn it down all the way and the engine cuts off, turn it back to halfway and try to start it again. Eventually you will keep it running with the choke all the way off and you're ready to go.

Again, read your owner's manual to find out the particulars of the engine you are maintaining. But they are all basically the same. The keys are to annually change the oil, the oil filter if it has one, the air filter if it has one, the spark plug/s (every couple of years is fine), grease any fittings, fill the tires with air and drain the gas for the winter or start the engine at least once a month to keep the gas from spoiling in the lines. It's that simple and once you learn on any small engine, you can do this maintenance on larger ones, even your tractor, car or truck. If you have a two-stroke engine, there's really not much maintenance. Just replace the spark plug every few years and make sure to drain the gas before storing for the winter.

An oil filter wrench is the perfect tool for unscrewing tight jar lids. You'll never have to ask a man to loosen a cap again!

Fixing a Flat Tire

Tires go flat on a farm quite frequently. Lots of random metal objects, some centuries old, seem to attach themselves to tires on a farm as if rubber were magnetic. They'll be lying in wait for decades for just that moment when you have three strips of grass left to cut with your lawn tractor. Then your tire springs a leak. As with an engine, both small tires and big tires work the same way and you should feel confident that once you learn the basics, you can fix either.

To help keep your tires in shape, you should have an electric air compressor on the farm. A standard compressor of 90 psi is suitable. Anything more powerful is meant to run air guns and other tools. A tire pressure gauge should go into each of your vehicles and in your tool box. It's hard to have too many as they are seemingly impossible to find when you need them. And finally, keep a tire plug kit so that you can fix any simple flat that hasn't produced a major tear in the tire.

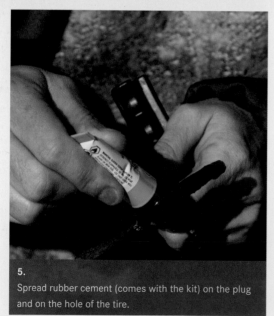

1.
Look for the source of the leak. If you can't find it, fill the tire with air and listen for the leak. If you still can't find it, fill a tub with water and put the tire in it. Rotate it in the water until you see bubbles and where they are escaping from. Remove the obstruction, if there is one, with a wrench.

2.
Use a tire plug kit and read the directions.

5.
Spread rubber cement (comes with the kit) on the plug and on the hole of the tire.

6.
Shove the plug hard into the hole until only an inch or so of either side of the plug is sticking out.

3.
Rough up and expand the hole in the tire with the plunger provided.

4.
Thread the rubber plug through the eyehole on the plunger.

7.
Turn the plunger 90 degrees and then pull it out with a jerk.

8.
Trim off the excess plug.

9.
Fill the tire with air to the proper pressure. Check with tire pressure gauge.

Removing the Tire

- First, try to find the source of the leak. Sometimes, you will not even need to remove the tire if you can just look closely and spot a nail or some other object in it. If you see where it is and it's easy to get to, don't remove the tire; proceed instead to the tire plug kit instructions.

- If you can't immediately locate the problem, you'll need to remove the tire. If you have a truck, you should buy a bottle jack. Standard truck jacks don't always do the trick. And you can use a bottle jack for your tractor or other equipment.

- Lots of times you'll be in a field and the jack will sink into the ground. So, you'll want to have blocks around—these are just 4 × 6 or 6 × 6-inch pieces of wood (buy a post at your home store and cut it into 2-foot lengths) that you can set the jack on to keep it from sinking. Blocks are also good to keep on hand because trucks and tractors are high off the ground and hard for a regular jack to reach from the ground.

- Set the parking brake. Make sure the truck is on level ground and the jack is as well. You can drive the truck a short distance on the rim if you have to get to a better spot.

- Loosen the lug nuts before jacking it up, but don't remove them all the way. Another tool you should have around is a large X-shaped lug nut wrench. You get a lot more leverage with it.

- Jack the truck up, putting the jack on the suspension or frame—find out from the owner's manual the best spot to place the jack. For a lawn tractor, you'll need to lift up the corner of the tractor with help from another person and slide a wood or concrete block under it to hold it up.

- Take the lug nuts off.

- At this point, you can always throw the tire in your car and take it to a mechanic to fix.

Fencing

Matching your fence to your needs is essential. There are many types of fencing and all have specific uses, from keeping animals in or out, to keeping neighbors in or out. Some fencing, like the new electric netting, is meant to be mobile and to move with grazing animals. Other fencing, like board fencing, is meant to hold large animals like horses without having sharp edges that might cause injury. Barbed-wire fencing should only be used to contain large cattle. Animals injure themselves on barbed wire. Cattle have very thick hides and are not as easily injured. A good way to educate yourself about fencing is to go to your local Tractor Supply or other farm supply store and look at all their fencing options. Ask questions and you can usually get a mini tutorial on how to install any type of fencing.

Installation of fencing is basically the same for any type you choose. First, you run a string line along the area where you want your fence. Then you measure and mark where the posts will go at the proper intervals for your type of fencing. Marking paint is ideal for this purpose. You then install your posts by either pounding them into the ground with a special tool or digging holes to insert them. When the posts are installed, you run or string your fencing and tighten it to the desired tension for the type of fence you're installing (you wouldn't tighten wood, of course). And lastly, you attach the fencing to the posts, either with nails, plastic or wire clips or metal fasteners, depending on the type of fence. Very important in the installation process is to reinforce your corners to the specifications of the fence you are installing. Fencing is not a technical undertaking, but it does require planning and is much easier with several warm bodies as help.

Electric Fencing

Electric fencing can be used to contain any animal. The smaller the animal you are keeping, the closer to the ground you will want to run your first line. And on the lower lines, you'll want to run them close enough together that whatever animal you might be keeping cannot squeeze through. You can also run a combination of electric and other fencing if you have animals that continually try to escape non-electric enclosures. And a line or two of electric fencing at the top of a garden fence will deter deer or predators from getting in. You can run electric fencing directly from an electrical outlet if you have one nearby. Or you can buy a solar unit that has a car battery with it to run fencing that is away from a power source. The size of the solar charger

is dependent on how much fencing you are powering. Your fencing dealer can help you determine the size you need.

PROS: Electric fencing is inexpensive and easily movable. It does not require a lot of grunt work to set up.

CONS: Electric fencing can fail if a branch falls on it or there's a power failure. It requires frequent checking and repair.

ANIMALS: Good for use with any type of animal. Cattle should have a supplemental restraint like barbed wire as electric fencing is very easy for them to break through if they are determined.

Electric Netting

Electric netting is now one of the most popular types of fencing for animal farmers. It's easy to install as it comes with stakes at the end of the posts that you just push into the ground. You make a circular enclosure with it around your animals. You can buy extra stakes to make sure it stands up well. It's best to use a solar-power source as the whole idea is that it's a highly mobile unit. The purpose is to protect your animals from predators while also being able to move your animals onto fresh pasture every few days.

PROS: easy to install and move; is effective for predator control.

CONS: is moderately expensive; wears out quickly if you move it often.

ANIMALS: best for smaller animals like chickens, sheep and goats.

High-tension wire and T-post

This is some of the easiest fencing to install and maintain. It also lasts for many years, even decades if installed properly. It's easy to repair and it is very safe for animals.

PROS: Easy to install, inexpensive, durable and safe for animals.

CONS: Only useful for larger animals as it's easy for smaller animals to slip through the space between lines.

ANIMALS: For use with large animals like horses, donkeys, llamas and cattle.

It's important to secure corners. The hardware for corners is available where you buy your fencing supplies and comes with directions.

Woven wire

For keeping small animals in and predators out, woven wire fencing works best. You can buy it in different sizes. The five-foot-tall woven wire fencing with four-inch-by-two-inch openings is what we use around the perimeter of our garden and we've never had a predator or deer make its way inside. It's also very safe as the openings are small enough that there's no room for animals to get stuck. It lasts for twenty to thirty years. You can attach it to t-posts or to wood posts. The t-posts last much longer and tend not to lean as easily. You install it the same way as the high-tension wire fencing. But you'll need to stretch and create the tension yourself. We attach one end of the fencing to the end post, unroll it and use a truck or tractor to pull the other end so that it's very tight. Then we attach the fence to the posts. You can also use a fence stretcher attached to your truck or tractor.

PROS: lasts a very long time and is very strong; keeps even the smallest animals in or out; good for predator control.

CONS: not aesthetically pleasing and will eventually rust; hard to repair once it's installed; is expensive for large areas.

ANIMALS: good for all types of animals. Not practical for large animals or spaces.

Fixed Panel

Fixed panel is similar to woven wire. But instead of flexible fencing that you unroll and stretch, you can buy panels generally in ten- or twelve-foot sections. They are stiff panels that you attach to any post you prefer to use. These are good to use with pigs as they are very safe for animals and very strong.

PROS: very safe for animals. Easy to install.

CONS: expensive; is easy for some animals to climb over.

ANIMALS: best for pigs and other animals that stay close to the ground and don't like to climb.

Black painted boards don't have to be painted as often.

Board fencing

Certainly the best looking fencing of all is board fencing. Who doesn't like the sight of a white board fence stretching into the distance? Board fence is ideal for horses as it's sturdy and very safe for animals. It will not keep predators or deer out. Also, the cost and maintenance is prohibitive for many people. Board fencing is the most labor intensive to install. If you install board fencing, the best thing to do is to rent a post-driver, not a post-hole digger. A post-driver is like a big hammer that will hammer the posts into the ground. An auger or post-hole digger will dig out a hole that requires you to reinforce at intervals with concrete. So a post-hole driver will save you that trouble.

PROS: safe for animals; aesthetically pleasing.

CONS: not effective for predator control; requires expensive and labor-intensive maintenance like painting.

ANIMALS: for use with large animals only, like horses, cattle and llamas.

Hand Tools

There are too many possible tools that you might have around a farm to list them all. You will certainly need common tools in various sizes like hammers, wrenches, screwdrivers, shovels, rakes and hoes. There are other tools described in detail in other parts of this book, like an air compressor and a fencing tool. I won't attempt to cover everything. But there are certain tools that I've found to be particularly useful in my farm work. Learning to use those tools has saved me time, labor and money. As important is learning how to sharpen tools with an edge. Most tools you use around plants, like hoes, shears and loppers, work best when sharp and clean.

Left to Right:

A harvest knife is essential for thick-stemmed vegetables like broccoli.

Pruning shears, needle nose flower shears and garden/vegetable shears

Shears, clippers, and harvesting tools

Look inside any gardening catalog and you will find numerous sizes and shapes for tools to prune, weed or harvest plants. Regular garden shears are a must, along with hand spades and hand cultivators. There are a few that I use very often and I find most effective: needle nose shears, garden shears, a Japanese digging tool and a harvest knife.

Shovel

Shovels are about the most common tool you will use. But there's now a shovel specifically designed for women. The Hershovel from Green Heron Tools (see Resources) has a large, wide handle that gives you more leverage with your upper body and an enlarged step that helps maximize the digging power of your body weight. Their website will guide you to fit the shovel to your specific needs and body type.

Left to Right:

No weed is safe from a Japanese weeding tool!

Heavy duty shears for harvesting herbs

The Hershovel in action.

Pick mattock

Another tool that isn't all that special until you have to dig a hole is the pick mattock. Most people reach immediately for a shovel when needing to dig a hole. But shovels get jammed on rocks and require a lot of back strain to move packed earth. The pick mattock is the best tool for digging. It has a sharp pick on one side that can crack through rock and the flat blade on the other side easily wrenches up dirt, clay and sod. They come in different sizes and weight to fit any body.

Straight crowbar or digging bar

Most crowbars are just a couple of feet long. But you can get a straight digging bar or straight crowbar that is four or five feet long. Why would you want a heavy metal bar that's so long? This tool is ideal for lifting heavy objects just a few inches or feet. The bar gives even a small person enough leverage to lift the corner of a thousand-pound tractor implement. You can use it to adjust implements when hooking them up to your tractor. Or you can lift the corner of a lawn tractor or pallet that you want to put up on concrete blocks to work on or get off wet ground. It can also be used to break up rocks at the bottom of a deep fence post. It's quite versatile for a rod of metal.

Hoes and precision hoes

A hoe is the everlasting symbol of tedious garden work; picture the elderly woman hunched over and hacking away. But technology in the form of precision hoes has made weeding almost pleasurable work that is much easier on your back. There are all kinds of these long-handled hoes: wire weeders, collinear hoes, trapezoid hoes and stirrup hoes. They are made with an ergonomic long-handle that is attached to the head of the tool at a proper angle so you can work it while standing upright, with a straight back.

Wheel hoe

A wheel hoe is much more than a hoe. It's a hand tool with a wheel attached to the front that comes with various attachments for working your planting rows. There are hoe attachments for weeds, cultivator attachments to loosen soil, plow attachments to create furrows and seeder attachments for planting. They aren't inexpensive, but they last a lifetime and are much more environmentally friendly than power tools.

Broadfork

A broadfork can serve as a manual tilling tool and is healthier for your soil and the living creatures in it than using a power tiller. It's used to loosen the soil in beds and work compost or fertilizer into the soil. There are two types: those with spike-like tines that are best on beds that have fairly loose soil already, and those that have blades that are used to break up really hard pan. Buy the best quality you can as cheap ones are easily broken. The all-metal broadfork we use is heavy, but it will never break. When using it to incorporate organic matter into the soil, make sure not to dig the deep layers up and mix them with the top layers of soil. You only want to work the tines into the soil and move them back and forth so the organic matter on top drops down the holes you are creating and you just loosen the soil up. Mixing the layers is damaging for the microorganisms in your soil.

An all-purpose fencing tool

1.
Use a spring-loaded post-hole driver to set the t-posts in the ground. They should be 1½–2 feet deep. Stand on a block or stepladder to get better leverage.

2.
Secure the tension cams to the last post with the fencing wire that is recommended to be 14 gauge or heavier. Use a joiner or crimping tool to secure it with the fencing wire.

5.
Use the special cam wrench to tighten the wire. It's best to have someone pull the wire away from you to provide some tension so that you can get it going. Once it begins to tighten it's much easier to crank. Tighten as tight as you can get it as the wire will stretch a bit.

6.
You can buy the post clips or they are usually given free when you buy the wire. Secure the wire to the post using your fencing tool, or you can also buy a special tool (shown) that is easy to use.

3.
Gripple brand wire joiners are the easiest and longest lasting way to splice and attach the wires to one another. You insert the fencing wire into each side. The wire easily slides in and then is near impossible to pull out. It costs more, but is more effective and easier to use than a crimping tool. And it's ideal for when a fence wire breaks and you need a quick way to resplice it.

4.
After securing the wire all the way at the other end with the Gripple joiner, unroll the wire to the cam you installed. Insert it into the hole in the middle.

7.
Twist the right side first and then the left.

8.
A T-post puller is used to get T-posts out of the ground if you ever need to move or replace them.

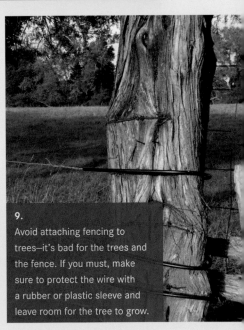

9.
Avoid attaching fencing to trees—it's bad for the trees and the fence. If you must, make sure to protect the wire with a rubber or plastic sleeve and leave room for the tree to grow.

Seeders

Seeding by hand is perfectly adequate for a small space. But once you begin to farm for market, you'll want to make seeding more efficient. A broadcast seeder is perfect for planting seed that can be sown directly on top of soil, like cover crop, over a large area. The seeds are held in a hopper and you turn a handle so the seeds shoot out and are broadcast evenly as you walk. A walk-behind row seeder is a more advanced tool and come in various styles, sizes and prices depending on your need. They generally run on wheels, dig a small furrow, inject the seeds at adjustable intervals, and then cover the seeds as you push the tool down the row. You can buy single-row or multi-row seeders and different attachments that fit the specific types of seeds you are planting.

Sharpening tools

Garden tools work best when clean and sharp. The best time to sharpen is at the end of the season when you're putting your tools away for winter. If you sharpen and oil them with a bit of mineral oil or silicon spray, your tools will stay rust free and be ready to go when the first weed breaks the soil in spring. But this of course is not always possible and sharpening in winter or early spring is perfectly acceptable. And sometimes, you need to sharpen just to get a tool to work properly again. Sharpening gardening tools is not an advanced art, although it can seem intimidating. Remember, you're not sharpening Excalibur or a samurai sword. It's just a gardening tool and you only need it to make your life easier. Before you begin, clean all the dirt from the metal and dry off. If there's a lot of rust to be removed, spray the metal with WD-40 or a similar loosening spray and let it soak for awhile. Then take a wire brush to it to remove the rust. It doesn't have to be spotless. Start with a coarse sharpener and scrape down the metal until smooth. Then use a fine sharpener to create a sharp edge. Find the angle of the blade and sharpen along the angle.

Hand Sharpener

Typically, these are called diamond hoes or diamond sharpeners. One side is coarse for removing loose metal and generally smoothing the blade. The other side is fine and finishes the sharpening. A hand sharpener is recommended for smaller tools like shears, but can be used on larger tools as well. Don't be afraid to really scrape it down with a back-and-forth motion using the coarse side. Then only scrape toward the tip of the blade when you switch over to the fine side. And it's always good to give the back side of the blade a few swipes to really hone the blade to a tip.

Rotary Tool Sharpener

This attachment fits onto the end of your drill. It has a perfect angle for most tools. Just run the tool along the blade a few times and you'll be sharp in no time. This is a rough sharpening tool for hoes and the like and doesn't have a fine attachment to really hone a blade like that of a knife.

Bench Grinder

A bench grinder is recommended for the biggest blades like brush shears. A bench grinder typically has coarse and fine stones. Be careful not to let the blade heat up on the stone too much or you will damage or break it. Use quick, even strokes to sharpen it up and use safety goggles.

Tractors

Tractors are wonderful tools on a farm. They are also expensive. You can get a good deal on an older tractor, but the older ones are more difficult to drive and don't usually have four-wheel drive. I use a tractor primarily to move mulch, dirt, or manure around the farm. So, while we have a large tractor, I could get by with one much smaller. The newer, four-wheel drive tractors with power steering are as easy to drive as a car and what I'd recommend for most women farmers. But it all depends on the amount of work you plan to do and the size of the equipment you plan to pull.

Three-Point Hitch and Power Takeoff

These are what separate a lawn tractor from a tractor. If you want to run such equipment as a tiller, brush hog, fence-post driver, or combine, you will need a tractor with a three-point

hitch and power takeoff. The three-point hitch consists of two side bars that attach to an implement, and the draw bar on top that attaches and adjusts to make the equipment level. The power takeoff is a power source that you hook up your implement to, to power it. For instance, it's what creates the power to turn the blade on your bush hog.

Front-End Loader

This is a bucket used for scooping and carrying, and is optional on the front of tractors. I strongly suggest that you buy a tractor that has one. Adding one later is a very expensive proposition. It's a very versatile attachment. Not only can you scoop dirt, rocks, or whatever can fit in the bucket, but you can also pick up heavy equipment by hooking a chain to it. You can also use it as a lift so you can raise another person in it to do work in a high spot, such as the top of a hoop house.

WOMAN-POWERED FARM

Tips for Buying a Tractor

- Buy as new as you can afford.

- Find your closest tractor repair shop and buy a brand that it services.

- Buy a tractor with four-wheel drive and power steering.

- Make sure the tractor has a modern roll bar.

- Buy the size that's right. Your tractor dealer can help you match the size to what type of farming you will do. And even if you don't buy from the dealer, you can learn a lot by talking to its salespeople.

- Try to buy a tractor that comes with various implements. If you buy used, you might get all the extra equipment you need in one package.

Useful Implements

- **Front-end loader:** scoops materials, lifts heavy equipment, and serves as a power ladder

- **Backhoe:** the single bucket on the back of a tractor that digs holes, trenches, and stumps

- **Shredder or bush hog:** what you will mow your pasture with

- **Plow:** There are various types of plows that are used to break ground in different ways.

- **Disk:** levels and breaks up the ground after it's been plowed

- **Blade:** for leveling driveways and plowing snow

- **Auger or post driver:** for installing fence posts

- **Tiller:** mixes up soil to prepare for planting and incorporating cover crops

- **Wood splitter:** for firewood

Walk-behind tiller

A tiller is a two-wheeled machine with tines underneath that mix up the soil. I use a tiller occasionally, but it has its drawbacks. It is good for tearing up weeds in looser soils. But it can create a hard pan underneath the mixed soil that is unhealthy for the roots of plants. I try to only use it when weeds in a bed become so bad that it's the only way to get them up without lots of manual labor. I also use it to mix in cover crops. But beware that tillers mix the layers of the soil and damage the delicate balance of your soil food web. It kills and disrupts the bacteria, fungi and microbes that live in your soil. So use a tiller sparingly.

Walk-behind tractor

A very good alternative to a full size tractor, especially for women who do not want to wrestle with large tractor implements, is the walk-behind tractor. A walk-behind tractor at first resembles a walk-behind tiller. And it does have a tiller attachment. But it is much more than that. Walk-behind tractors were designed by Italian farmers and are ideal for small farms and farms in mountainous or hilly terrain. They come with almost all the implements that a riding tractor has: mowers of all kinds for cutting grass and brush, flail mowers for knocking down cover crops, rotary plows for mixing cover crops into the soil, wood chippers, snow blowers, seeders, pressure washers, tillers, harrows, spaders. The only thing a walk-behind tractor doesn't have that a riding tractor does that I can think of is a front-end loader. But the implements are smaller and easier to attach and use.

There are two brands of walk-behind tractor: BCS and Grillo. The best dealer we know of was one of the first to offer these tractors in the U.S.—Earth Tools in Owenton, Kentucky (see Resources). They are also the experts in repairing these tractors. Walk-behind tractors are not cheap, but you can have one with several implements that you will need for under $10,000.

We have had both a small BCS walk-behind tractor and now a larger Grillo one. These tools have made our two-acre organic market garden possible. The size of our walk-behind tractor fits well with the scale of our farm: it allows us to have permanent beds and paths, makes it easy to have a compact and diverse farm layout, and operates in our two high tunnels. The flail mower has opened up a whole new world of incorporating cover crops into our crop rotation.

–CONNIE LEMLEY
Cedar Ring Greens, Frankfort, KY

Trucks

My truck's name is Jack. I've had him for 13 years. He used to drive me to and from school, but now he's retired to the farm and he hauls firewood, mulch, lumber, and trash. I can't imagine having a farm without a truck. There's just too much to haul to get by without one.

You don't need a new one. Your farm truck is going to get beat up. It'll get scratched up by brush. It'll have dents all along its bed from various pieces of wood, stone, or metal that get thrown in the back. The seats and carpets will have dirt and dried mud caked on them from your boots and from your dog's riding shotgun. Don't buy new—so you don't notice its slow deterioration quite as much.

Everyone has her truck preferences, but the keys to look for in your farm truck are:

➡ Does it have four-wheel drive? I strongly suggest that it should.

➡ Is the bed 6 feet or longer? It should be if you want to haul lumber, your lawn tractor, or loads of any consequence.

➡ If you're going to pull a trailer, is it the right size for the trailer you'll be pulling?

➡ Does it have a name? If not, name it.

Things to Keep in Your Farm Truck

➡ **Tie downs:** Get some nylon tie downs that ratchet down to secure loads in the bed of the truck.

➡ **Tarp:** to cover loads.

➡ **Bungee cords** of various sizes

➡ **Bottle jack:** in case of a flat

➡ **Large X-shaped lug wrench**

➡ **Wood block** to set a bottle jack on if needed.

➡ **Adjustable wrench:** helpful when a battery is dead and you need to remove it

➡ **Jumper cables**

➡ **Tire pressure gauge**

➡ **Long length of nylon rope or a chain with hooks:** to pull brush, stumps or other stuff you might need to drag a short distance. Also useful to tie off an area of broken fencing until you can return to fix it.

➡ **1 quart each** of oil, brake fluid and power steering fluid.

GROWING ON THE FARM

CHAPTER   SEVEN

Vegetable gardening was a priority for me when we first moved onto our farm property. While I was growing up, my family had always had a kitchen garden, so throughout my childhood I was accustomed to enjoying fresh and delicious vegetables from spring through fall. For ten years prior to starting a farm, I'd been living the city life of rental rooms and apartments. Imagining my own home always included a kitchen garden and flowerbeds and only for my own personal enjoyment. A garden was as much a part of my mental image of "home" as the bedrooms or bathroom. But I did not start out looking to be a farmer.

I finally had a place of my own now that we lived in Virginia. I had land on which to grow things, and lived in a place where things seemed to *like* growing. I already had several wonderful gardening books. *The Garden Primer*, by Barbara Damrosch, was a mainstay and is still an inspiration for me, and I had been given a useful *Country Skills* book by one of my students in California as a "good-bye" gift. My previous gardening had taken place over ten years earlier, and had always been in the role of helper; I had a bit of rusty hands-on knowledge, but needed to acquaint myself with the basics of starting from scratch. As the mid-Atlantic climate was completely new to me, I checked out piles of library books and scoured them for information. I also found the agricultural extension agency's website and it has been an invaluable reference for information about growing things in my specific part of the country.

I began a garden journal, using a hardbound blank book, and recorded my climate zone and the average first and last

frost dates. I printed out and taped in a schedule of planting and harvesting dates for vegetables. And I sketched out a plan for the garden, carefully marking out its size, and the placement and size of the beds. I still have that gardening journal. It is much the worse for wear and is stuffed full of other notes, handouts, and loose sheets of information. I've been tempted a couple of times to get more organized, and once even started sorting and moving things from the journal into a binder. Now I have a binder with a few things in it, along with my scruffy, bulging journal. The main point is, have a place to keep your gardening notes. You'll want to be reminded of what you did the previous season and how it worked. Hopefully you jotted down helpful suggestions to yourself (Net the dahlias *early*!!!) and maybe noted when the first Japanese beetle made its appearance, and where the ladybugs seem to be hanging out. These notes build on themselves and become more meaningful every year.

As a woman, what are the most important things to consider when setting up a garden? Time is, of course, a precious resource. You may have children that need to be taken to school and sports. You may be homeschooling your children and need to set time aside in the day for inside studies. You may even have an off farm job. Labor is also key. As we discussed in the "Healthy Farmer" chapter, it's very important to take care of your body. So, when setting up your garden, you'll want to minimize the distances you have to walk and plan your garden for a size that is manageable for you and whoever might be helping you. Money would be the other key element. Do you have the resources for greenhouses, wooden raised beds, predator-control fencing? All of these factors should be considered when planning your garden. Obviously, the most efficient use of all three—time, labor, and money—is the goal.

Before we discuss setting up a garden, we should consider the size. There's a difference between a kitchen garden, which is used as more of a supplement to your cooking, and a farm garden planted to produce significant yield that you might live off entirely or sell what you grow for profit. You can be more free-form and whimsical with a kitchen garden because it is smaller and you are not dependent on producing a specific volume of produce for market or restaurant sales. But we are going to consider a farm garden, or market garden, as there are many books about keeping a "backyard" garden. There are a few principles besides time, labor, and money in considering a farm garden. And the principles would apply the same to whatever size garden you create.

The Five Keys for Maximizing your Growing Efforts

➡ **Location:** Where you locate your garden in relation to the sun and your living conditions is of primary importance. Both the location and orientation of your garden can greatly affect its success and the wear and tear on the farmer. The farther you have to travel to get to your garden, the more labor and time you use. And if you have to get in a car or truck to drive to your garden, then you're spending money as well.

➡ **Water:** Rain, of course, is the ideal water source. If you have about 1 inch of rain per week, you may not need to water on your own. And a cool running spring or river from which you can divert water to a cistern for storage or directly to your plants is nice. But most people don't have this luxury. So, the only supplemental watering option that is both sustainable and healthy for your plants is drip irrigation. Overhead watering invites fungal diseases and water is wasted in vast amounts through evaporation. Drip irrigation delivers deep watering directly to the roots of your plants with minimal waste. It's also easy to use with permanent raised beds and is really the only way to farm sustainably unless you have adequate rainfall each year where you live. The amount you have to water on your own involves time, labor, and money.

- **Raised bed growing:** There are various forms of raised beds, for perennial growing and annual growing. You can also grow in raised beds under cold frames in cold parts of the year. There are other forms of planting, including conventional row planting, but permanent raised beds (those that you build once and never move or disturb greatly) are the best way to maximize your soil health because you're limiting soil disturbances to protect the microorganisms, fungi, worms, and other beneficial creatures in the earth. By using them, you are preventing erosion of the topsoil. Permanent raised beds save time, labor, and money in the long run, as you only have to do the initial building of them once. For these reasons, it's the best option for the small farm garden.

- **Uniformity:** If you create your garden plots and rows the same length and width, you can use all of your row covers, drip irrigation, landscape fabric, etc., interchangeably, which saves time, labor, and money.

- **Soil food web:** Plant/vegetable/flower gardeners are often called dirt farmers. This is because their primary goal is healthy soil. What they grow on it is the by-product. Mulches, cover crops, and compost are the fundamentals of soil health and the keys to creating a healthy soil food web.

Choosing a Site

The first thing to do when planning a garden is to choose a site with fairly level ground in full sun. Vegetable and flower gardening require full sun, at least 6 to 8 hours in the middle of the day, with southern exposure. You can grow plants with exposures to the east, west, and north (in order of effectiveness), but you will not get an entire day of sun and it will limit what you can grow and how well.

When selecting a site for your garden, access for yourself and your tools, machinery, and supplies is a high priority. If your garden is a pain to get to, you're not going to get there very often. And if it is troublesome to transport your tools and supplies, they won't get there, either. If you have a choice, make your garden as accessible as possible. Make it easy to enter and exit with good gates and space to drive a truck into, or right up to. If you'll be mowing between or around rows or beds, make sure you plan for space to maneuver the mower. Anything you can do to make the routine tasks of gardening easier will ensure that these tasks are more likely to get done. Think about where you will dump your weeds and debris, where you will locate your composting system, where you will store your hand tools and wheelbarrow. And just in case you want to, will you have space to expand your gardening area in the future?

The soil is the great connector of lives, the source and destination of all. It is the healer and restorer and resurrector, by which disease passes into health, age into youth, death into life. Without proper care for it we can have no community, because without proper care for it we can have no life.

–WENDELL BERRY
The Unsettling of America: Culture and Agriculture

It would be great if the site with the best sun exposure and ease of access also had the best soil on your property. Now that we've lived on this piece of land for 12 years, I'm not sure there really was a spot that had any better soil than another—it is all clay and rocks. You work with what you have.

Laying Out Your Garden

Planning out a structure for your garden is not only a creative labor of love, it can make a huge difference in how comfortable it is to work in your garden, and even how you grow and work within it. Over the course of a day, you will potentially walk miles through your garden. Positioning its structures and the layout of your rows efficiently can greatly reduce your labor and time.

When deciding how to position your garden rows, first consider the types of crops you will be growing; the light and soil needs of the plants will determine where to plant them. You also want to consider how you, their caretaker, will be moving through the garden—what shape and amount of space you will need to tend your plants. Make sure your paths are wide enough to fit a wheelbarrow or garden cart, and that your rows or beds are not so wide that you can't reach to the middle to plant, harvest, and cultivate. If you can't reach from one side of a bed all the way to the other or you cannot straddle the bed, then it will add extra work for you.

It is also important to think about the placement of the perennials relative to the annuals in your garden. Perennials need a fixed spot and won't be a participant in any crop rotation schedule you have planned. If they spend most of the summer as a tall and ungainly presence, such as asparagus, place them toward the back corner of your garden—easy to get to in spring, but later camouflaged by robust summer and fall crops.

Annuals are plants that complete their entire life cycle in one season. Annuals must be planted from seed every year. Their goal is to grow as quickly as possible and produce as many seeds as they can before they die or are killed off by frost. In milder climates, annuals can self seed and thus may *appear* to come back every year.

Perennials are plants that may die back each fall (such as peonies), but their root systems stay alive and produce new growth in the spring. To stay healthy over the years, perennials require maintenance, such as pruning, fertilizing, and sometimes dividing the root system.

Raised beds or rows? A small to medium kitchen garden lends itself to the use of raised beds contained by wood. Raised beds that measure 4 × 4 or 4 × 8 feet are easy to build and convenient to place for easy planting, harvesting, and crop rotation. Raised beds also provide a neat and tidy appearance. Make sure to use untreated wood when constructing wooden raised beds. Pressure-treated wood has chemicals that will leach into your soil.

If you are creating a production garden for market or wholesale produce, you will most likely want to plant in long rows. This way, your seeding and transplanting can be done with a wheel seeder or by tractor; drip irrigation can be installed in long, continuous strips; and you'll be able to till and fertilize the entire area by machine. Long, straight rows also make harvesting a large volume of crops much more efficient.

The most efficient and also the best method for building soil is by planting in permanent raised bed rows. You build standardized rows that you enrich every year with cover cropping, the addition of compost, and building soil in the pathways. You can minimize or do away entirely with tilling, which will protect the microorganisms in your soil. You can work these rows with a walk-behind tractor and over time build very deep and rich soil that produces much more food per square foot than a conventional tractor row. See page 190 for more on how to build permanent raised beds.

Efficiency and Uniformity

As most women now getting into farming are interested in small gardens and not hundreds of acres of conventional crops, efficiency and uniformity of your garden should be of the utmost importance. If you maximize your space, you can grow more food or other crops per foot than many a conventional grower. This will maximize your efforts onto a smaller space, which saves time, labor, and money. Growing in this way, or biointensively, will increase your ability to generate income while decreasing the wear and tear on your body; you'll be working on a smaller space that takes less heavy lifting to farm.

TIPS FOR THE MOST EFFICIENT GARDEN LAYOUT

- Regardless of the size of your garden, from a quarter-acre to many acres, divide your growing area into at least two and preferably more rectangular plots of the same size. This will make it easier to plan crop rotations.

- Make all your beds 30 inches wide. Most tools and walk-behind equipment are made for this width. Also, it's easy to straddle the beds while you work. Make all your beds the exact same length. This makes it easy to maintain drip irrigation, row covers, tarps, and landscape fabric.

When you pull it up, you know that any piece can be used on any bed without labeling it.

- The pathway widths are dependent on the type of plants you're growing. Vegetables can be grown using 18-inch pathways. For our garden of cut flowers, we need more space and have our pathways set at 36- or 48-inch widths so the flowers can spread out. Regardless, keep a uniform width.

(adapted from *The Market Gardener,* by Jean-Martin Fortier)

Biointensive agriculture is an organic agricultural system that has been shown to result in maximum yields from the minimum area of land, while simultaneously improving and maintaining the fertility of the soil. It is particularly designed for the small-scale grower. Plants are grown very close together in deep beds as the roots will grow straight down. This maximizes the space you're growing in and keeps weeds down.

Also take into consideration how big the plants are going to grow and how much support they will need. Tomatoes are a good example of a plant that starts off quite small and ends up taking up a space of 4 feet wide by 6 feet high and needing a sturdy support system. Will your crop grow upward (beans, small cucumbers), or must it be allowed to sprawl along the ground (melons, squash)? When we expanded our vegetable garden, we used the cedar posts embedded for the original fence on that side to construct permanent trellises. We alternate growing pickling cucumbers and sugar snap peas on each half in the spring, and in midsummer, plant edible and ornamental flowering vines, such as hyacinth bean and climbing nasturtium.

If your garden is small and you plan to use a bed for more than one crop in a season (succession cropping), you'll need to plan ahead to get your first crop into the ground, so it will finish in time for your second crop to be planted and come to maturity before the end of the season. Be sure to follow crop rotation guidelines as well as enrich your soil with organic fertilizers and compost between plantings—this will keep your soil healthy and discourage pests.

Water

The next consideration is water. If you're lucky, you have some choice in the water source you will use to irrigate your garden. Our property has a deep, drilled well, and that is my main source of intentional watering. Additionally, we built a creek-fed pond on our property a couple of years after moving in, and although it does get low when we experience several weeks with little to no rain, this source of surface water has never dried up and provides necessary water for wildlife and our grazing animals. However, anytime I can supplement with harvested rainwater—in my case, a 100-gallon do-it-yourself rain barrel system—I do. Whenever possible I use this water for my hand watering (seedlings, flowers and herbs in pots, even new plantings if there aren't too many). Seedlings and new plantings require initial watering because their root systems haven't yet grown to extend deeply into the ground. Once your plants are established, the general

WOMAN-POWERED FARM

rule of thumb is to forgo watering if your garden is receiving an average of an inch of rain a week.

Having grown up in New Mexico, I have a very clear sense of how limited and valuable water is. Groundwater is water that's accumulated over thousands of years. It's stored in the Earth and takes a long time to replenish. I wish I didn't have to use any groundwater at all. However, because occasional drought conditions make it necessary for me to use groundwater, it is essential that it be delivered to my plants in the most efficient and effective way possible. To do this, I use a drip irrigation system. In this manner the water goes right where it is needed, the plant roots, with little waste through evaporation or run-off. Drip irrigation is fairly easy to install and can be tailored to any size bed or horizontal planting system. Several sources for drip irrigation system supplies and components are listed in Resources.

If your garden is small and you have the time to spend, watering with the traditional garden hose is simple and easy; just take care not to overwater. When I first started out, I only watered using the hose. But I was also teaching at the time and it became impossible for me to spend the hours necessary each day to water my growing garden. A slow, sustained watering ensures that the water you are adding at the top of the soil has the opportunity to meet up with the plant roots several inches below the surface. You may also consider buying a soaker hose at your local garden supply center. This is a hose with tiny holes along its length and that is closed at its end. Once the water pressure builds up in the hose, the pressure forces the water out of the tiny holes to water the ground along the entire length of the hose. This method will save you some time, and is a similar principle, on a smaller scale, to drip irrigation. But soaker hoses get clogged up easily, are generally short in length, and don't seem to last more than one season.

Now that you've got a site for your garden, it's time to start thinking dirty. Or at least thinking about your soil.

Soil Food Web

If you have attended conferences on farming and agriculture, you may have heard the expression "Feed the soil, not the plant." This is one of the mantras of organic and sustainable agriculture, and with good reason. But what exactly should we be "feeding" our soil? The soil-testing kits available at garden centers are a handy and easy way to test for nutrients and pH levels in different parts of your garden. Your county agricultural agency will test soil samples that you mail to them, and send you an analysis of the nutrient levels in your soil with recommendations of what they should be.

An incredible diversity of organisms make up the **soil food web**. They range in size from the tiniest one-celled bacteria, algae, fungi, and protozoa, to the more complex nematodes and microarthropods, to the visible earthworms, insects, small vertebrates, and plants. As these organisms eat, grow, and move through the soil, they make it possible to have clean water, clean air, healthy plants, and moderated water flow.[8]

Soil Testing, pH, and Basic Nutrients

Managing pH and nutrient levels is important, especially for encouraging the vigor of certain crops with special needs (such as blueberries). Plants need soil with a specific pH range; without it, they are unable to take up nutrients from the soil. The pH is the level of acidity in your soil. A low pH reading means your soil is more acidic. A high reading means it's alkaline. Most plants grow in a pH range of 6.0 to 7.0. So, test your pH after every season with a pH kit or meter. All gardening books, testing kits, plant tags at stores, and many online resources can tell you the ideal range for every specific plant imaginable. If your pH needs to be raised,

The Soil Food Web

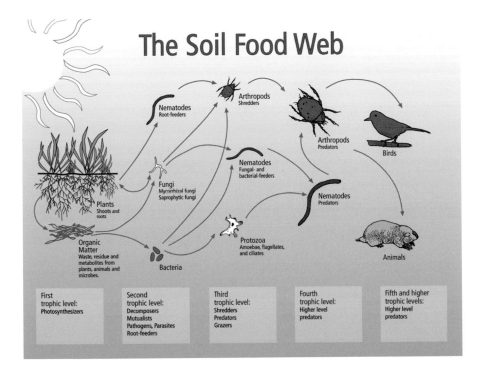

First trophic level:
Photosynthesizers

Second trophic level:
Decomposers
Mutualists
Pathogens, Parasites
Root-feeders

Third trophic level:
Shredders
Predators
Grazers

Fourth trophic level:
Higher level predators

Fifth and higher trophic levels:
Higher level predators

buy some lime (the rock kind, not the fruit kind) from your local garden center and follow the directions on how much to add to your soil to raise the pH to the amount you need. If the pH needs to be lowered, use greensand or sulfur, also available at the garden store and with specific directions for how much to use.

Testing your soil for other nutrients every few years will let you know what other nutrients might be depleted. Contact your local agricultural extension agency to obtain the testing kit with instructions. You'll be testing the three main ingredients for plant health—nitrogen, phosphorous, and potash, or N-P-K. A professional test from your ag agency will also include many micronutrients. You can use a store-bought kit (pictured), but it won't be as accurate.

There's an organic fix to any nutrient deficiency. Be careful of the recommendations of your agricultural extension office.

These easy-to-use kits are ideal for testing pH levels in the soil. You can also test for other nutrients, but a test from your agricultural extension agency is more reliable.

The workers are trained to use chemical nitrogen to fix just about everything. It's best to just look at what nutrients are missing and go in search for the specific organic fertilizer or amendment that is missing. Your local organic gardening center can help you.

The long-term health of your soil depends not on simply treating symptoms shown by certain plants, but on supporting the soil food web. As I've learned more about the soil food web, I've become fascinated by this incredibly complex and diverse living system. In the book *Teaming with Microbes*, the authors show that "By using techniques that employ soil food web science as you garden, you can at least reduce and at best eliminate the need for fertilizers, herbicides, fungicides, and pesticides. You can improve degraded soils and return them to usefulness." Adding organic matter to your soil is the key to a healthy soil ecosystem. You can accomplish this with compost, compost tea, organic fertilizers, mulches, and cover crops.

Compost

Your own homemade compost is a wonderful way to feed your soil. In nature, the organisms that live in the soil are constantly decomposing organic matter into simpler compounds, but the process can be very slow (think of that wonderful, cushiony soil covered in pine needles, deep in the woods). As the microorganisms break down the organic matter, they release nutrients in a form that the plants can absorb. Composting is simply a method of speeding up the natural decomposition of organic materials.

Creating your own compost can be as simple or as complex as you would like to make it. The basic tenet is that the microorganisms need both carbon and nitrogen food sources. Carbon sources are called *browns* due to their brownish color and provide the energy source that microbes need to break down organic matter. Nitrogen sources are called *greens*, although

At any given time one pile is usually finishing up, or ready to be used (Left), a second pile is about midway through the process of decomposing (Middle), and the third pile is the new pile (Right).

not all nitrogen sources are green, and generally contain more moisture than browns do. Nitrogen sources provide the protein that allows the organisms to grow and reproduce. Some commonly used browns are straw, leaves, wood chips, and sawdust. Commonly used greens include grass clippings, manure, vegetative kitchen scraps, and garden debris, such as rotted or damaged vegetables and plant parts. When building a compost pile, supply carbon and nitrogen at a rate of 30:1 to create an atmosphere in which the pile will heat up and decompose more speedily. This means you will need 30 times more browns (carbon source) than greens (nitrogen source).

Never add dog, cat, or human feces to your compost pile. These contain dangerous pathogens that may not be killed during the composting process. Also avoid animal kitchen scraps, which may attract rats.

I've used a three-pile system for about 10 years now, and I build it using a "lasagna layering" method. My first layer is one of browns. I usually put any stalky garden debris at the bottom to provide air pockets, as well as straw and leaves, to form about a 10–inch-high pile, then add a layer of greens (usually a 5-gallon bucket full of saved-up vegetarian kitchen waste) to make about a 2-inch-high layer. I continue alternating layers, ending with a brown layer. I don't weigh and measure to achieve the 30:1 ratio; I approximate and it has worked fine for me.

I use mostly straw and leaves as my source of carbon, and keep those piles right next to my composting area so that I can quickly and easily layer them with my greens—mostly vegetative kitchen scraps and donkey manure—to make a new pile. I collect kitchen scraps in a small bucket under my sink. When that is full, I dump them in 5-gallon bucket with a tight lid, in my back-yard. I keep several of these 5-gallon buckets, and when they are full, it's time to create a new compost pile.

When building a new pile it is necessary to moisten the pile because the dry browns are so much more numerous than the greens. Remember, my leaves and straw are approximately 30 times more numerous in my pile than my manure and kitchen scraps. The microorganisms need that carbon energy to do their work. Also, depending on the temperature outside, you may need to moisten throughout the composting process. I cover my piles with black plastic or an old tarp to retain the heat and moisture (this helps keep the microorganisms active) and protect the piles from the sun and extremes of weather, to prevent the nutrients from leaching out. I turn them every couple of weeks or so until the compost has achieved a consistent crumbly dark brown appearance. Finished compost should smell like fresh earth and be free of any large chunks of material. My piles are usually 4 × 4 × 4 feet, and because I don't turn them terribly often, it takes about three to six months for them to break down fully.

There are wonderful books written solely on the topic of compost and composting methods. I've included a few of these in Resources (see page 331). There are also many reputable sites on the Internet that cover composting from the most basic to the most scientific and elaborate. It seems that every farmer and gardener has a favorite recipe for mixing greens and browns. Compost is one of my favorite things. I urge you to explore different methods of composting and play around until you find a system that works for you.

Basic Compost Setup

- A 4 × 4-foot area lined by fencing or some other material, such as pallets

- Black plastic to cover the pile and protect it from the sun (which kills microorganisms)

- Water to keep it moist but not soggy

- Enough brown material (leaves, straw) and green material (kitchen scraps, grass clippings, manure) to layer a pile that is 3–4 feet high and consisting of three-quarters brown material in height to one-quarter green material. Layer it evenly.

While I was growing up, we always had a compost pile. It was a fenced area about 4 × 4 feet, where we simply dumped our kitchen and yard waste and let it do its thing. This wasn't the fastest way to create compost, but it did provide an environmentally friendly way to recycle this waste. This is sometimes referred to as "lazy" composting because you really don't do anything but wait for the materials to break down naturally—which they will eventually.

It is difficult to make enough compost to meet the needs of an entire garden or farm. And if you factor in the heavy lifting, the turning and creation of large compost piles is not an ideal chore.

One way to get around this is to find an organic compost supplier and have the compost delivered to your farm. It's not cheap, but I've never felt like I didn't get my money's worth once you factor in all the physical effort it takes to make even a small pile of compost. Making your own compost is a great way to use your leftover kitchen scraps and lawn clippings, but there is a much more efficient and less physical way to treat your garden with all the benefits of compost without a massive physical effort—compost tea.

Compost Tea

An efficient and very effective way to make a little compost go a long way is to make *compost tea*. This saves time, labor, and money. Compost tea is a fantastic way to feed your soil food web as it contains a diverse and myriad number of living microorganisms that are multiplied by the process of oxygenated brewing. To be clear, compost tea is not the liquid that sometimes leaks out from the bottom of the compost pile (what is called compost leachate). This liquid is not a good source of any nutrients. Compost tea is made by soaking finished compost in water. The most effective way to make the tea is to also aerate the water with an aquarium pump, so as to add oxygen to aid in the growth of beneficial microorganisms in the compost.

The process and materials for making aerated compost tea follow. One teaspoon of regular compost has a bacterial population of 1 billion. And 1 teaspoon of compost tea contains a bacterial population of 4 billion.[9] Bacteria and fungi are needed in your soil as they carry nutrients to your plants and their roots. A good compost tea has all the key nutrients your plants need, including N-P-K and many micronutrients; it also contains living bacteria and fungi and other microorganisms that make those nutrients available to your plants. So, you're really supercharging your compost by aerating it and making tea out of it. Just one 5-gallon bucket of compost tea can inoculate an entire acre of your garden with beneficial microbes and fungi. That's a whole lot easier than spreading a ton or more of regular compost that it would take to do the same thing. So, make a small pile of compost from your kitchen scraps, garden waste, and animal manure that is well cured and then turn it into hundreds of gallons of compost tea. In this way, you'll save yourself countless hours of physical labor.

The quality of your compost that you make your tea with is very important. Remember, the process of brewing the tea will multiply whatever pathogens are in the compost, even the bad stuff, such as *E. coli*. It is essential that your compost be fully cured before using it to make compost tea. And you should never use manure in your compost tea unless it's been fully composted in a regular compost pile.

There's also a distinction between *aerated* tea and *anaerobic* tea. Anaerobic tea is made by just putting compost into water and letting it sit for a while. It does not have the benefit of oxygen to grow all the good stuff. This process takes a long time and is dangerous, as it breeds pathogens and has little microbial activity in the end. You might as well just put the dried compost on your plants. The tea we are talking about here is aerated, which is the process of adding oxygen to the water to create an active tea of living organisms.

There are endless recipes for compost tea; my favorite follows. You can include all kinds of additives that feed either the bacteria in your compost or the fungus. Bacteria-dominated tea is ideal for vegetables and annuals. Fungi-dominated tea is ideal for perennials. If you are interested in the science behind this, I'd recommend the book *Teaming with Microbes* (see Resources). But you can make compost tea with only finished compost and molasses. Anything else you put in there is just a bonus.

Without getting lost in the scientific weeds, I recommend sticking to a balanced compost tea that has both beneficial bacteria and fungus.

BASIC COMPOST TEA

1.
Place your homemade compost in the cheesecloth sack.

4.
Place your aerator stones in the bucket and plug them in.

5.
Add the molasses to feed the microorganisms.

2.

At this point you may add any further dry amendments you like to the sack (bat guano, worm casings, kelp, fertilizer, etc.).

3.

Tie a knot in the sack and place it in the 5-gallon bucket of fresh nonchlorinated water to about 3 inches from the top. If all you have is chlorinated water, fill the bucket and let it sit for several hours to allow the chlorine to evaporate.

6.

Allow the compost tea to brew while covered for 24 hours.

7.

Give it the smell test—if it smells like sweet earth, it's done. If there's any hint of sourness, brew it a little longer.

8.

Use immediately, as the living microorganisms will perish soon. If you need to keep it a little longer, add more molasses and you can keep brewing for about 12 more hours. The longer you wait, the less potent your tea will be, however.

9.

If you will use it in a sprayer, you'll want to pour it through a paint strainer first.

Basic Compost Tea Recipe (limit your ingredients to 4 cups or less)

Ingredients

2 to 3 cups finished compost (or worm castings if you don't have compost)

⅛ cup bat guano

½ cup worm castings (if you haven't used it already in place of compost)

¼ cup kelp meal

¼ cup balanced organic fertilizer

2 tablespoons unsulfured molasses

Supplies

Cheesecloth sack

1 (5-gallon) food-grade bucket (ensure that it is food grade or you will have bits of plastic and chemicals added to your tea)

Aerator stones

Fish tank air pump—the more power the better

Vinyl tubing

1 (5-gallon) paint strainer (optional)

A standard rate of diluting the compost tea is one part tea to three parts nonchlorinated water, although you can go to as high as 1:25. If you divide your compost tea into three 5-gallon buckets and fill them up with water, you'll have 15 gallons of super compost tea. The tea is most immediately effective when applied with a fine, powerful sprayer onto the surface and underside of plant leaves. This is because the microbes adhere to the plants and offer the nutrients directly into them, while giving them some protection against insects. However, it can also be used as a soil drench—applied directly to your garden beds by hand or with a hose end sprayer—to inoculate and fortify your soil.

A cover crop of buckwheat and rye.

Cover Crops

Cover crops are another method of adding organic matter to your soil. Cover crops are grasses, legumes, or small grains that are grown specifically to help maintain and build soil fertility and productivity. They are used not only to improve soil fertility and health, but also to help prevent erosion, to suppress weeds, and even to aerate heavier soils.

There is an astonishing selection of cover crops, and it can be overwhelming if you are new to cover crops. Luckily, there are some very helpful seed catalogs out there, and you can make your cover cropping adventure as simple or complicated as you like. What type of cover crop you decide to grow depends mainly on your purpose in growing the crop (mulch, soil fertility, erosion

Farm Seed Comparison Chart

SEED TYPE	SOWING SEASON	MINIMUM GERM. TEMP.	HARDINESS ZONE	GROWTH RATE	SOW PER 1,000 SQ. FT.
Alfalfa, Summer	early spring–late summer	45°F/7°C	frost sensitive	fast	½ lb.
Barley	early spring–summer	38°F/3°C	7	fast	2 lb.
Buckwheat	spring - summer	50°F/10°C	frost sensitive	fast	2-3 lb.
Clover, Crimson	anytime	45°F/7°C	7	medium	⅔ lb.
Clover, Mammoth Red	anytime	41°F/5°C	4	fast	½ lb.
Clover, Medium Red	anytime	41°F/5°C	4	medium	½ lb.
Clover, New Zealand White	spring–summer	40°F/4°C	4	slow	¼ lb.
Clover, Sweet	spring–summer	42°F/6°C	4	medium	½ lb.
Cowpeas	spring–summer	58°F/14°C	frost sensitive	fast	2 lb.
Mangels	spring–summer	41°F/5°C	frost sensitive	medium	¼ lb.
Manure Mix, Fall Green	summer–fall	45°F/7°C	see components	medium	1½ lb.
Manure Mix, Spring Green	spring–summer	38°F/3°C	see components	medium	5 lb.
Millet, Pearl	summer	60°F/16°C	frost sensitive	fast	¼ lb.
Mustards	spring–summer	40°F/4°C	7	fast	¼ lb.
Oats	spring–summer	38°F/3°C	8	medium	4 lb.
Oats, Hulless	spring	38°F/3°C	8	medium	4 lb.
Peas, Field	spring or fall	41°F/5°C	7	fast	3 lb.
Radish, Oilseed	late summer	45°F/7°C	6	fast	1 lb.
Rape, Dwarf Essex	spring–summer	41°F/5°C	7	fast	1 lb.
Rye, Winter	anytime (fall for grain)	34°F/1°C	3	medium	4 lb.
Ryegrass	anytime	40°F/4°C	6	fast	1 lb.
Soybeans	spring–summer	60°F/16°C	frost sensitive	fast	4 lb.
sudangrass	early summer	65°F/18°C	frost sensitive	fast	1 lb.
Sunflower	spring	70°F/21°C	frost sensitive	medium	1,500 seeds
Sunn Hemp	summer	72°F/22°C	9	medium	1lb.
Turnips	spring or late summer	45°F/7°C	6	fast	¼ lb.
Vetch, Chickling	spring–summer	45°F/7°C	8	medium	2 lb.
Vetch, Hairy	anytime	60°F/16°C	4	slow	1 lb.
Wheat, Spring	early spring	38°F/3°C	7	fast	4 lb.

SOW PER ACRE	SEEDING DEPTH	NITROGEN FIXATION	BEES/ BENEFICIAL INSECTS	COMPACTION CONTROL	EROSION CONTROL (COVER CROP)	WEED SUPPRESSION	PEST MANAGE-MENT	GREEN MANURE	FORAGE	BIOMASS (ORGANIC MATTER)
15–25 lb.	¼–½"	yes	yes	×		×			×	×
80–125 lb.	¾–2"				×	×			×		×
50–90 lb.	½–1½"		yes			×			×		
22–30 lb.	¼–½"	yes	yes		×	×			×	×	
5–15 lb.	¼–½"	yes	yes	×	×	×			×	×	
5–15 lb.	¼–½"	yes	yes	×	×	×			×	×	
5–15 lb.	¼–½"	yes	2nd year		×	×			×	×	
10–20 lb.	¼–1"	yes	2nd year	×	×	×			×	×	×
70–120 lb.	1–1½"	yes							×		
10 lb.	½–1½"									×	
50 lb.	½–1½"	yes	yes			×			×		×
200 lb.	½–1½"	yes				×			×		×
6–10 lb.	½–1"				×	×			×		×
5–12 lb.	¼–¾"		yes		×	×	×		×	×	×
110–140 lb.	½–1½"				×	×			×		×
110–140 lb.	½–1½"				×	×			×		×
120 lb.	1½–3"	yes				×			×		
10–20 lb.	¼–½"			×	×	×					
5–15 lb	¼–¾"							×		×	
60–120 lb.	¾–2"				×	×			×		
20–30 lb.	0–½"				×	×			×		×
150 lb.	1"	yes									
30–40 lb.	½–1½"			×	×	×			×	×	
20,000 seeds	½–1"		yes								×
30–40 lb.	½–1"	yes						×	×		
8 lb.	½"									×	
70 lb.	1"	yes			×				×		
25–40 lb.	½–1½"	yes	2nd year		×				×		
60–150 lb.	½–1½"				×	×					

prevention, weed suppression, soil aeration, or a combination of these), your climate (some crops grow in hot or cool weather), and the time of year you intend to plant the crop. There are other considerations, but these are the basics.

The main cover crops I grow are annual rye for weed suppression (because it grows quickly), and buckwheat and soybeans for nitrogen replenishment. The seed for these is readily available at farm supply stores and through mail order, and fairly inexpensive to buy in bulk. The buckwheat has the added benefits of being a tasty meal for my chickens and also producing a lovely white flower that provides a nectar source for my honey bees. The soybeans are a multipurpose crop as well. They fix the nitrogen in the soil and then provide us with delicious edamame (the steamed beans you can find at Japanese restaurants) to eat and sell at the market.

Cover crops add organic matter and nutrients to the soil when you till them, or work them into the soil with a broad fork, before they go to seed. Cover crops are best used mixed together so that the varied depths of their roots aerate your soil deeper.

Mulches

Mulching is another essential strategy in taking care of your garden and your soil. During the growing season, mulch provides a protective layer to guard against weeds, and also helps keep plant roots cool. In the winter, mulch provides protective insulation from the cold and can make the difference between a crop's overwintering or perishing. Using organic matter for mulching increases soil tilth and nutrition as it slowly decomposes. Mulch also makes your garden look tidy and cared for.

My favorite mulch (and a great compost addition) is autumn leaves. There is so much nutrition in leaves and they are there for the raking. To help the leaves break down more quickly, they can be shredded with a lawn mower, but I often simply spread them over a bed in the fall and then weigh them down with a layer of llama manure. In the spring, what is left of this mixture gets turned into the soil. This has been my standard winter mulching for my asparagus and dahlias and has worked well. If pine needles are readily available in your area, they make attractive mulch for plants that like a more acidic soil. Grass clippings and straw are traditional mulches for suppressing weeds throughout the growing season. Both are readily available and inexpensive.

In some cases it may be more effective to use a nonorganic mulch, such as black plastic or landscape fabric. We use both of these in different instances. We grow our strawberries in black plastic that has a small hole cut for each plant. I grow many of my cut flowers in rows covered in landscape fabric with a hole burned for each small seedling. When I plant out the flower seedlings, they are simply too small to shade out the weed seeds that are inevitably present in my soil, and would be completely covered by any organic mulch I could apply. Also, many growers, especially in high tunnels or greenhouses, use landscape fabric in the pathways. This is an expensive option, but certainly effective in keeping the weeds from multiplying. You can buy biodegradable landscape fabric, but it doesn't last long and then doesn't break down in the soil as quickly as is usually desirable. But it's certainly a more environmentally friendly option.

A few cautions regarding mulch: because mulch retains moisture, it can cause plants to rot. There are certain plants that don't like to be mulched, and, as is the case with my tiny flower seedlings, some mulch can smother new plantings. Mulch is attractive to burrowing rodents and can also attract slugs and snails. You'll want to experiment with different types of mulches to find out what works best for you and your specific gardening needs. Beware of dyed mulch. It has chemicals in it that will leach into your soil.

Building Soil in the Pathways

Your paths are your access to your garden beds and work areas. They need to be wide enough for you to maneuver a garden cart or wheelbarrow throughout the growing season. Our pathways are 4 feet wide because flowers tend to spill out of their beds into the pathways. It may seem tempting to use less space for your paths so that you have more ground for growing, but remember that your small seedlings will grow and sprawl. One season of having to navigate through your garden as if it were a jungle is all you need to realize and respect the value of the space devoted to your garden paths. But if you are growing in a small space, your space for growing should definitely take precedent over your pathways. When constructing the paths in our garden, we use a combination of laying down cardboard first, then covering the cardboard with a couple of inches of hardwood chips. You can find hardwood (or softwood if that's all that's available) from your local tree trimming service. Many times the workers will come and dump loads for free if they are working in your area. This creates an excellent weed barrier and provides attractive access to the garden rows. Additionally, as we walk on the paths and the woodchips and cardboard break down over the course of a couple of growing seasons, they eventually create a nice, rich hummus that we then till into the pathway soil underneath. This makes a rich, unused soil that we shovel onto the raised beds and build them higher. Then we start over again and lay cardboard and woodchips in the pathway to create more soil, which takes another season or two. Combined with compost, this is an ideal way to create a steady stream of soil to refresh your garden beds.

BUILDING SOIL AND BUILDING PATHS

1.
After removing grass with tiller or digging by hand, spread some manure. It can be fresh. This step isn't necessary, but will create much more nutrient-rich soil.

2.
Lay down cardboard that is free of plastic or laminate.

3.
Spread wood chips or mulch on top of the cardboard as deep as you can. Enough to keep the cardboard from showing through when you walk on it

4.
In about a year, you'll be able to dig out the rich soil under the mulch in the pathway and spread it on your raised beds.

RACHEL WILLIAMSON

Fairweather Farm

Afton, Virginia

Rachel Williamson made a radical change in her life. Over the course of two weeks, she decided to stay home and work the land where she owned her home. She quit her full-time job, ordered seeds for everything she ever wanted to grow, and made a ridiculous number of Excel spreadsheets for planting, growing, and harvesting specifications of everything she was going to do in the coming year. In the end, the best gift she gave herself was not deciding what her business was going to be before she launched into it. Instead she spent the first year trying out everything she was interested in—cut flowers, dried flowers, basket making, wedding flower arrangements, herbal teas, culinary spice blends, fresh herbs, garlic, and probably some other things she doesn't even remember.

She tried wholesale and direct marketing, craft shows, and Internet sales. She had about a dozen different businesses that first year, and just had to keep reminding herself that this was just a trial—that by pursuing multiple ideas, she wouldn't able to do any of them justice. It turns out that this was just what she needed to do. She currently runs the mixed herb, dried flower, and culinary spice business she calls Fairweather Farm.

"My goal is to lead a life that nourishes myself, my community, and the land I live on. Farming is a strong way to work toward that goal because the effects of my actions are so readily obvious, tangible, and concrete. I'm getting constant feedback on my conduct. There is no question that I am fully responsible for how my day feels to me, what I have to offer at market at the end of the week, and the health and well-being of the ecosystem in my care. The perfect trifecta is when land, self, and community are in a mutually strengthening relationship. What makes me happy is also beneficial for the land and my community. When I feed the soil, I produce better crops for my community and I reduce stress on myself. When sales are up because other people are enjoying what I do, it makes it more possible for me to continue living the life I love and working the soil in a wholesome way."

Over the years, Rachel had apprenticed at half a dozen places, learning everything from large-scale vegetable production to small-scale intensive gardening, mostly working for room and board. She never worked on a farm that does exactly what she is doing, but did pick up many invaluable skills. Even if the end product was completely different, most of the skills were transferable. She thinks it was just as important to learn all the things that she wanted to do differently as it was to learn the skills she wanted to replicate. This background experience gave her the confidence and skills to jump right in.

"I'm so glad I went that route instead of writing a business plan and locking myself into one specific vision of what my farm was going to produce, because it left me open to really find out what I enjoyed doing and what there was a strong market to support. Thankfully they ended up being the same thing! I look back at that now and wonder where on earth I got that kind of confidence, but wherever the faith came from, I'm glad it did," she says. Somehow she still ended up in the black that first year. She's made enough to buy her own piece of land next door to her brother, another farmer. And she hopes to relocate her current operation there eventually.

Fairweather Farm is a joyfully abundant acre of land in Afton County, Virginia. Rachel describes it as, "A little farm growing a lot of herb teas, culinary spices, and poplar bark baskets. I like to think that when you taste my teas and spices, it's a burst of flavor undeniably here and now, something you'll never experience from the grocery store."

She has booths at the two largest farmers' markets in her area, the Charlottesville City Market and the Nelson County Farmers' Market, and travels all over the country to participate in craft shows and holiday markets. In her booth she has a sign with four pictures of her farm mounted on a beautiful piece of weathered wood and the captions read, "One

WOMAN-POWERED FARM

acre, One woman, 100% Homegrown, 100% Handmade." That pretty much sums it up.

"It's not that I feel particularly separate from men in the business, but I do feel a unique kinship with other women who are living similar lives. Mothers and fathers both love their children but there's something undeniably different about mothers, and being a woman farmer is a similar distinction," she says.

Here are some of the attitudes and skills that Rachel has found most important in beginning a small farm and growing a business:

- Love learning, because there's no end to what you do not know.
- Be okay with not knowing how to do things—suddenly you're going to have to be a marketing expert, graphic designer, soil scientist, accountant, licensing whiz, chef, mechanic, construction worker, entomologist, and about a hundred different things. There's no way you're good at all of that in the beginning.

- Pay attention. Respond to market feedback, what's happening in your garden, and how your body is doing.
- Be a realistic optimist. Not everything is going to go as planned, but you can almost always glean something positive from anything that comes your way.
- Love what you do. If you aren't enjoying yourself and passionate about what you're doing, there's no way you're going to want to work as hard as farming necessitates. She says that it's true she works 12 hours a day seven days a week, but she's doing what she loves 100 percent of the time. You can't beat that. She doesn't have to wait until five o'clock to find a little time for hobbies, work out, or get some nutritious food. She's doing what really feeds her—mind, body, and soul—all the time.

See page 331 for Rachel's wonderful list of resources.

Crop Rotation

Crop rotation is another great friend of the organic and soil-conscious gardener. *Crop rotation* simply refers to the practice of not growing the same crop in the same location year after year. The main reasons to practice crop rotation are soil (plant) health, and pest and disease control. The basis of all crop rotation schedules is plant families. Crop rotation can seem a bit complicated, but it doesn't have to be. Draw a rough sketch of your garden and make a few copies; this helps with planning out your plantings, and is a good reference for where you've planted certain crops in the past. On my garden diagram I have my original vegetable garden beds identified by letters, and my long rows by numbers. My hoop houses get a separate diagram with numbered rows. Of course, all of this is kept in my garden journal.

Being aware of which plants belong to which family, and what nutrients this family depletes and leaves behind, will help you to plan out your crop rotation for the season and the years to come. For example, the Fabaceae family (I call it the "bean" family) grow fine in average soil but leave behind traces of nitrogen, making that spot a good place for planting "leafy greens" next year (Brassicaceae family). Another reason to rotate crops is to prevent soil-borne disease. When a crop is moved to a different bed, the soil pathogens that were attracted to that crop no longer have a suitable host. This same principle works for pest control. Many pests overwinter or lay their eggs in the soil where their favorite food source grows. If you move that crop, the pest will have to waste energy searching for its favorite food and is less likely to survive and thrive.

FAMILY NAME	CROPS
Apiaceae	carrots, parsnips, celery, dill, chervil, cilantro, parsley, caraway, fennel
Asteraceae *also known as* Compositae	lettuce, endive, escarole, radicchio, dandelion, Jerusalem artichoke, artichoke, safflower, chicory, tarragon, chamomile, echinacea, sunflowers
Brassicaceae *also known as* Cruciferae	cabbage, cauliflower, broccoli, horseradish, kohlrabi, kale, Brussels sprouts, turnips, Chinese cabbage, radishes, rapeseed, mustard, collards, watercress, pak choi, bok choi, rutabaga
Chenopodiaceae	spinach, beets and sugar beets, chard
Cucurbitaceae	cucumbers, melons, summer and winter squash, pumpkins, gourds
Ericaceae	blueberries, cranberries
Fabaceae *also known as* Luguminosae	Beans, peas, lentils, peanuts, hairy vetch, vetches, alfalfa, clovers, cowpeas
Lamiaceae	lavender, basil, marjoram, oregano, rosemary, sage, thyme, mint, catnip
Liliaceae	asparagus, onions, leeks, chives, garlic, shallots
Poaceae	corn, wheat, barley, oats, rice, millet, rye, ryegrass, sorghum, Sudan grass, fescue, timothy
Polygonaceae	buckwheat, rhubarb
Rosaceae	apples, peaches, apricots, nectarines, plums, strawberries, blackberries, raspberries, pears, cherries
Solanaceae	peppers (bell and chile), tomatoes, potatoes, eggplant, tobacco, tomatillos

MAKING PERMANENT RAISED BEDS

1.
Mow or weed-whack down all the grass and weeds.

2.
Till up the entire area, both the pathways and the row to be built.

6.
Work the amendments and compost in with a broad fork

7.
Smooth with a rake.

3.
Shovel the pathway dirt onto the bed to build it up.

4.
Add any organic fertilizer or amendments your soil test shows you need.

5.
Add several inches of compost or fresh organic topsoil.

8.
Lay out your drip tape for your drip irrigation and secure it with landscape pins.

9.
Cover with landscape fabric or plant without it.

Crop rotation is most effective when a bed is planted with a specific family once every four years. A four-quadrant crop rotation system is a good way to begin crop rotation and learn about plant families. Many gardening books provide detailed plans for this type of system. The basics are that you divide your garden into four quadrants, then assign one of the four largest vegetable families (Fabaceae [beans], Brassicaceae [leafy greens], Cucurbitaceae [squash], and Solanaceae [tomatoes]) to each quadrant. You then rotate each family through each quadrant over a four-year period.

What type of crops you grow will depend on your purpose in growing, the climate and altitude in which you live, and how much time you plan to spend in the garden. Once again, your state's agricultural extension agency will have a wealth of information pertaining to your particular area.

If this is your first garden, remember, start small. Don't overwhelm yourself with too many different types of vegetables or herbs. Have fun—experiment and play! Take the time to observe your garden as it changes throughout the seasons. Your vigilance and care are the most important tools for making your garden a healthy, happy place.

Seeds and Seed Starting

As you begin to make your decisions about what crops your will grow in your garden, it is important to know your Plant Hardiness Zone.

The Plant Hardiness Zone Map is a guide to help gardeners determine which plants will thrive in their part of the country. The USDA planting zones are regions defined by a 10°F difference in the average annual minimum temperature. The higher the number, the warmer the temperatures for gardening are in those planting zones. It is standard practice for seed dealers and nurseries to label their products according to their USDA

USDA Plant Hardiness Zone Map

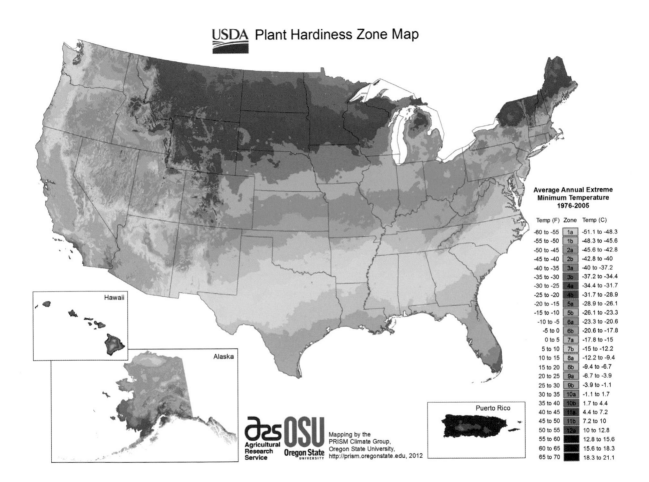

Hawaii

Alaska

Puerto Rico

Average Annual Extreme Minimum Temperature 1976-2005

Temp (F)	Zone	Temp (C)
-60 to -55	1a	-51.1 to -48.3
-55 to -50	1b	-48.3 to -45.6
-50 to -45	2a	-45.6 to -42.8
-45 to -40	2b	-42.8 to -40
-40 to -35	3a	-40 to -37.2
-35 to -30	3b	-37.2 to -34.4
-30 to -25	4a	-34.4 to -31.7
-25 to -20	4b	-31.7 to -28.9
-20 to -15	5a	-28.9 to -26.1
-15 to -10	5b	-26.1 to -23.3
-10 to -5	6a	-23.3 to -20.6
-5 to 0	6b	-20.6 to -17.8
0 to 5	7a	-17.8 to -15
5 to 10	7b	-15 to -12.2
10 to 15	8a	-12.2 to -9.4
15 to 20	8b	-9.4 to -6.7
20 to 25	9a	-6.7 to -3.9
25 to 30	9b	-3.9 to -1.1
30 to 35	10a	-1.1 to 1.7
35 to 40	10b	1.7 to 4.4
40 to 45	11a	4.4 to 7.2
45 to 50	11b	7.2 to 10
50 to 55	12a	10 to 12.8
55 to 60		12.8 to 15.6
60 to 65		15.6 to 18.3
65 to 70		18.3 to 21.1

Agricultural Research Service

OSU Oregon State UNIVERSITY

Mapping by the PRISM Climate Group, Oregon State University, http://prism.oregonstate.edu, 2012 planting zones, thus indicating the planting zones in which you'll be most successful at growing those particular plants.

It is a good idea to make a list of the vegetables, herbs, and flowers you would like to grow, and how much of each variety you plan to produce. Look through several seed catalogs and take note of different seed varieties, prices, and plant requirements. I usually order seeds from about five different companies each year, mainly because I want certain varieties (I get my New Mexico green chile seeds from Seeds of Change in Santa Fe), and I purchase seeds from my local seed producer, Southern Exposure Seed Exchange, at the local shops in town.

Most likely, you will begin perusing, or devouring, your seed catalogs in October when they begin to arrive in the mail. I allow myself to go crazy and mark with a red pen everything I think I might want and fold over the corners of the pages. I then go back, with my list of what I actually need, and my budget in mind, and winnow down my choices. Many seed catalogs offer descriptions of the growing needs of their seeds. Typically they include the hardiness zone for the plant, its water, light, soil, and nutrient requirements; its resistance and susceptibility to certain diseases and pests; and its growth habit. I really like a few seed catalogs because of the detailed information they provide about growing their seeds. For example, Johnny's Seeds also provides a germination rate for its seeds, the number of days to germination as well as harvest, and whether the seed does better if started inside or direct sowed outside. I keep the seed catalogs around to refer to throughout the year.

Seed Terms

HEIRLOOM Open-pollinated varieties that either predate or are unaltered by modern breeding work

OPEN-POLLINATED Varieties that result from pollination by insects, wind, self-pollination, or other natural forms of pollination and produce seed that is genetically "true to type," meaning that the seed will result in a plant similar to the parent

HYBRID Results from the deliberate crossing of two different parent varieties from the same species. "F1" refers to first-generation offspring from these two distinct parent varieties. Plant breeders began producing hybrids as a way of combining the best traits of separate varieties into one. Hybrid varieties are usually bred to offer greater disease resistance, vigor, and uniformity than open-pollinated or heirloom varieties.

GENETICALLY MODIFIED ORGANISM (GMO) Refers to varieties that contain a genetic trait not normally occurring in the plant's DNA. Genes from one species are artificially implanted into the DNA of another species, the seeds of which contain combined genetics that would not exist in nature. GMO seeds are not allowed in organic practices.

ORGANIC Seeds are grown strictly without the use of synthetic fertilizers and pesticides, sewage sludge, irradiation, or genetic engineering. Seeds labeled as organic must be grown and processed in accordance with the USDA's National Organic Program (NOP).

My local agricultural cooperative extension agency's website was most helpful when I began my vegetable gardening in Virginia. I found comprehensive and detailed information on vegetables specifically recommended for my region of the state, complete with planting and harvesting dates. Getting your plants started and planted at the right time is an important part of having a successful garden season.

The websites of seed and plant companies are another great resource to assist you in your gardening. Peaceful Valley (www.groworganic.com) has how-to videos as well as articles on many relevant topics. Johnnyseeds.com has a "grower's library" that includes downloadable planting charts to help you plan out your season of plantings. Be sure to explore these sites as they are reputable and free.

You don't have to order seeds from a catalog, but catalogs are a good source for learning what is available and getting a sense of pricing. You can purchase seeds as well as transplants, or starts, at your local gardening venue. If you are buying starts, I recommend purchasing them from local farmers. Most starts at garden centers have been grown in huge quantities and may have traveled hundreds of miles before arriving at the shop. You don't know who actually grew them or their growing conditions. Buying from a local farmer means you can talk to the person who actually grew the plant and get any information you need right from the source. Also, you are supporting your local community and economy. Buy fresh, buy local.

Many vegetables (especially root vegetables, such as carrots and beets) and flowers do well when direct seeded in the garden. Greens (spinach, lettuce, mustard, etc.), cucumbers and squash also germinate and grow well when direct seeded. Sunflowers and most wildflower varieties (bachelor's buttons, love-in-a-mist, larkspur, etc.) are also happy to be direct seeded. Other vegetable and flower crops need to be planted as seed-

lings or starts. These are available from local farmers at your farmers' market, local greenhouses, or nurseries, or at the garden centers of most home improvement stores. Because they take a while to grow, and they prefer warm soil and temperatures, tomatoes and peppers are good examples of these crops. If you have the inclination and the space, you may want to start many of your vegetables and flowers from seed on heat mats and under grow lights inside.

Seed Starting

Starting your own seeds doesn't have to be expensive or complicated.

There are good reasons for starting your own seedlings.

◗ You can get a head start on the growing season.

◗ Growing your own plants from seed gives you a greater selection of varieties from which to choose.

◗ You can control the quality of materials you use to start your plants. By selecting a high-quality seed-starting medium (growing mix), gentle and organic fertilizer, and keeping your seed starting containers and growing environment clean and sanitary, your self-started seedlings will be stronger and more resistant to disease than those you purchase from a garden center.

◗ The joy of watching those little seedlings burst through the soil and nurturing them into plants is one of the greatest pleasures of gardening.

◗ After the initial investment in supplies, you will be saving money—especially when you consider the higher quality and diverse selection of your home-grown plants—and you can sell any surplus seedlings at your farmers' market.

Basic materials to use and where to get them:

- Shelving (I use wire racks purchased from a home improvement store, but you can make your own out of boards and cinderblocks or other available materials.)

- Fluorescent lights (home improvement store)

- Automatic timer (gardening supply store or online gardening company)

- Heat mats (gardening supply store or online gardening company)

- Seed-starting medium (Seek out local farmers who are selling their own high-quality mixes, or purchase from a local garden center.)

- Seedling trays and pots (garden center or online, or repurpose such containers as large spinach packages and yogurt cups)

- Organic seedling fertilizer (garden center or online)

To start your seeds, you need soil, moisture, heat, and light.

◆ Moisten your planting mix before filling your containers.

◆ You can start seeds in just about any container as long as it has drainage holes. The four- or six-cell containers that are commonly sold in garden centers are handy because they are of uniform shape and designed to fit into the standard 11 × 21.37 × 2.44-inch trays. You'll want to put your containers or cell packs in a tray so that the water has a place to drain.

◆ Fill your container about three-quarters full of planting mix and lightly pack it in.

◆ Follow the directions on your seed packet for sowing your seeds. Some seeds will need to be covered with a fine layer of soil; others should not be covered. Spray your new plantings with a fine mist of water.

◆ Label your seed starts, using wooden or plastic tags and a permanent/waterproof marker. Record your plantings in your garden journal or spreadsheet. Include the date, variety, number of seeds planted, and any other details you want to monitor.

◆ In most cases you will need to supply supplementary heat for your seedlings for them to germinate. Seedling heat mats are readily available at garden centers and online. To help keep the soil from drying out, you can cover your tray with a "humidity dome"—a plastic cover made to fit the seedling tray.

◆ Your seedlings will need a bright and continuous light source. I use broad-spectrum fluorescent shop lights attached to wire shelving with metal chain, so that I can raise the light as the plants grow. The light should be a few inches from the top of the seedlings. To simulate a natural growth cycle, turn off the lights at night. You can set up a timer to do this automatically.

◆ The heat mat, humidity dome, and light create a mini-greenhouse atmosphere. When your seeds germinate, the first "leaves" on your seedling are actually the food source from the seed itself and are called cotyledons. The first true leaves will look different from the cotyledons. Once your seedling germinates, you want to remove the humidity dome to provide air circulation and keep the surface soil from remaining too moist.

→ Depending on the size of your container, once your seedling has three sets of true leaves, it can be transplanted into a larger container. When the majority of the seeds have germinated, note the date in your records, along with an estimate of what percent of the seeds you planted actually germinated. You can compare this data to the information on your seed packet.

→ Keep your seedlings healthy by making sure they have adequate air circulation, and allow the seedlings to dry out slightly (not completely!) between waterings. Use a gentle organic seedling fertilizer sparingly.

→ Follow the directions on your seed packet or cultivation sheet for planting your seedlings outside—temperature, spacing, depth of planting, light, and soil considerations.

USING SOIL BLOCKERS TO START SEEDS

As an alternative, or in addition, to using the plastic cell contain-ers available in gardening centers, you can start your seeds in soil blocks. Soil blocks are compressed blocks of soil that stand closely together without needing any container. The blocks are formed with a specific tool, a soil blocker, using a moist seed starting mix. I didn't start using soil blocks until about four years ago, and I wasn't thrilled with them on my first attempt. I did more research because I liked the advantages that using soil blocks provided.

Advantages of Soil Blockers

- They eliminate, or cut down, on the purchasing, cleaning, and storing of plastic containers.

- Blockers use the seed starting space, including the heat mat space, more efficiently.They create stronger seedling roots by providing the seedlings with more oxygen and less root disturbance—this reduces transplant shock when you move your plants out to the garden. Instead of circling the pot, roots stop at the air barrier between blocks.

- It is easier to start a large amount of seeds at a time.

The reason my initial attempts at making soil blocks were unsatisfactory was that I was using a seed starting medium that did not bind together very well. Consequently, my soil blocks came out of the blocker only partially formed, or they broke apart quickly. In recent years, soil blockers have gained in popularity, and several of the seed companies that sell soil blockers also offer prepared soil mixes to use with the blockers, as well as provide recipes for making a good soil starting mix on your own. Trying to keep things as low maintenance and close to home as possible, I found a great organic potting soil at my local organic gardening center. It is a variety without perlite, which turned out to be the perfect medium for my soil blocking needs. The perlite caused the soil to crumble and not stick together. The soil without it retains moisture, so is great for retaining the form of the blocks, and also has great drainage, essential for preventing seedling rot. Additionally, it has plenty of organic nutrients, so I rarely need to fertilize my seedlings before I transplant them to the garden.

Planting Out Your Seedlings

You've created your rows, added organic matter to the soil, and it is time to put your seedlings into the garden. Following your garden plan, load up your little plants onto trays and carry them out to the garden. Use a cart if you will be planting several trays, and bring your trowel and a full watering can. The soil in your row should be moist, either from rain or a recent watering. Follow the spacing instructions on your seed packet. Use your trowel to make a series of holes within arm's reach. Take your cell pack or other container and, cupping the top of the plant with your hand, tip the container upside down, and squeeze gently on the sides of the container while easing the plant out into your palm. If you have used soil blocks to grow your seedlings, pick up a block with your fingers. Place the little plant into the hole slightly below the level of the row. If your seedling has grown leggy (has a longer, somewhat spindly stem), you can plant it with the leggy stem an inch or two below the row level. Firm the soil around the plant. Plant to fill the remaining holes within arm's reach, then water in each plant—water slowly until the soil around each new planting is saturated. Move down your row and repeat until all of your seedlings have been planted and watered in. You will need to hand water your new plantings until

1.
Pour your soil mix into your bucket or tub and add water to moisten. Let the mix sit for a couple of hours or so to absorb the water and check the consistency of the mix. When you scoop up a handful and squeeze gently, it should hold together with no crumbling. I've heard the consistency referred to as peanut butter, cooked oatmeal, and slurry. After you try making a few blocks, you'll quickly figure out how moist your mix needs to be. The level of mix in your container needs to be about twice the height of the blocker you are using. I use a 2-gallon bucket when I make mini blocks (¾ inch), and a 5-gallon bucket when I make 2-inch blocks. If you plan on making lots of soil blocks, you'll want to use a tub so that you can have more prepared soil mix to work with

2.
Push the soil blocker firmly into the mix with a twisting motion. Your goal is to fill the squares of the blocker snugly.

5.
I use cafeteria trays to hold the soil blocks. I can get 180–200 mini blocks on one tray. Continue to fill your blocker and eject your soil blocks onto the tray.

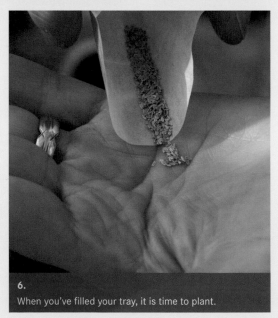

6.
When you've filled your tray, it is time to plant.

3.
When you pull the blocker out, turn it upside down to see whether all the squares have been filled completely. I often have to scoop up more mix and pack in onto the blocker, then use my hand or the side of the bucket to scrape away the excess.

4.
Place the soil blocker at one corner of your tray and depress the plunger to eject the blocks. Lift the blocker away in a straight up, smooth motion. It may seem awkward at first, but as you continue you'll get the hang of it.

7.
You'll notice that each block has an indentation on the top in which to place the seed. Place one seed in each block. Unless your seed packet says otherwise, you do not need to cover the seed with soil.

8.
Use a fine mist sprayer to moisten the blocks, being careful not to dislodge the seeds. Do not let the blocks dry out. The blocks along the edges of the tray will be the first to show signs of dryness. Once the seed germinates and begins to grow—developing a root system—you can water with a heavier spray and apply small amounts of water to the tray. Be careful not to overwater. Keep the blocks moist, but not wet (think of a wrung-out wash cloth, holding moisture, but not soggy or dripping).

their root systems develop enough for them to reach out and capture water from the surrounding soil.

Be attentive to the support needs of your plants. Not all plants need support, but those that do benefit from getting that support early. Place your cages around your tomatoes as soon as you plant them. If your crop requires a trellis, be sure to put that in place before you plant the crop. Put horizontal support netting on your flower rows before the plants start to flop over and sprawl out of their rows.

Harvest time will come before you know it, and it will keep on coming. Each crop is different, but in general, it is best to harvest early, especially if the days are hot. A fresh and stress-free plant means a harvest that will come off the plant looking better and keeping longer. Whenever possible, don't harvest if plants are wet from dew or rain because this can spread disease. Keep your crops in a dark, cool place until you are ready to take them to market. Just as with your other garden tasks, you will develop a system for harvesting. Utilize carts when you can and bring everything you need with you so that you don't have to make as many trips out of the garden. Utilize kneepads and a stool to protect your joints and back. Select the proper tool for harvesting a particular vegetable, and let yourself try new methods.

STARTING SEEDS OUTDOORS

Being able to direct seed a crop outdoors is a time and labor saver. Your seed catalog or packet will usually indicate whether that seed should be started indoors or can be direct sown outside. Remember, the seed packets are guides; ask other local growers about their planting habits, and experiment for yourself.

If you have the option, try to plant after a recent rain so that your soil has some moisture. Prepare your row for planting by raking the soil surface so that it is smooth and even. Remove any rocks or clumps of debris. Use a string line—a piece of string

stretched out from one end of the row to the other between two stakes—to keep your row straight. You can use a hoe to make furrows down the length of the row, drop your seeds in by hand, and use the hoe to cover the seed. If there is no rain in the forecast, you will want to lightly water the row (but don't water too much or you'll wash away your seeds).

While this method works fine, and I often plant my sunflowers this way, using a wheel seeder makes completing the task quicker and more efficient, and works well for long, very level rows. A wheel seeder is a mechanical, walk-behind planter. Wheel seeders come with different plates that have been specially designed for the spacing and seed size of different crops.

1.
Remove the sod and grass using a pick mattock.

2.
Dig out the top four inches of soil and pile up on a tarp.

5.
Fill the cold frame with the topsoil on the tarp.

6.
Add afour inches of compost, or peat moss mixed with organic fertilizer

3.
Smooth out the remaining dirt and cover with cardboard.

4.
Position the cold frame over the bed.

7.
Plant and water.
Note: using cardboard will keep weeds out, but will limit you to initially growing shallow-rooted plants like lettuces and herbs. After the first crop, you can begin planting more deep-rooted plants.

8.
Enjoy the fruits of your labor.

As you push the seeder down the row, the seeds drop into the holes in the plate and are positioned in a furrow created by the seeder, and then soil is pulled over the seed by a chain or disk. You need level ground that is not too soft to get a nice, even distribution of seeds.

Broadcast seeders are used to spread large quantities of seed shallowly, over a broad area. They are also used to spread garden fertilizers, lime, and sand. The smaller size of broadcast seeder is handheld, but there are larger walk behind models, as well as very large tractor pulled varieties. For my purposes, the handheld size works well. I mostly use my broadcast seeder for spreading cover crop seeds.

Frost Protection and Season Extension

Be prepared to protect your crops from a late spring or early fall frost. From repurposed plant pots to large structures, there are many different ways to protect your plants, depending on the size of the crop you are protecting and how far temperatures will drop.

Save pots from larger landscape plants and collect them from friends and neighbors. These can be used as a quick shelter for small, individual plants when an unexpected frost threatens. Simply invert a pot over each plant at dusk. If it is windy, weight the pot down by placing a stone on top. Remove the pots in the morning. I've done this with eggplants, tomatoes, peppers, and basil.

Row covers are great protection for a whole row of crops. The fabric is available in different sizes and temperature ratings. The lightest protects down to 28°F and allows 70 percent light transmission. The greater the frost protection, the less light transmission you will have. Like the insect barrier floating row cover, these row covers can be stored and used year after year.

I use a medium-grade row cover (down to 26°F) to protect cold-tolerant plants that I set out early in the spring, such as snapdragons. Frost-protective row covers are most often spread over a succession of wire (number 9, available at home improvement and garden centers) or PVC hoops pushed into the ground at approximately 2 to 2 ½-foot intervals, and weighted down on either side.

Depending on how you anchor your fabric, access to your crop can be problematic. Some farmers bury the ends in the soil; others use landscape staples (large metal pins). I've tried the staples, but don't like how they leave holes in the cloth. My solution has been to roll the excess fabric on each side onto a 2 x 2-inch board as if a bolt of fabric, and weight down the board with rocks at various intervals. In this way I can lift the row cover to work in my row by lifting the board and its fabric up and over the hoops to the other side. It is also easy to remove the row cover to put away for the next use. If you intend to protect your crop through colder temperatures, snow, and ice, you need to construct a sturdier structure—a low tunnel.

Low tunnels are the next step in providing protection for crops as winter moves in and the temperatures stay low. While the fabric row covers and wire hoops are fine for frost protection, they will give way under snow or ice. You can still use the wire hoops, but will need to fortify them by threading heavy twine or wire between each hoop, looped around the hoop, and on down the row, finally pulled taut and attached to a stake at either end of the row. One length of twine or wire across the top, and one along the upper arch of the hoops (on each side) will provide support and strength. This structure can then be covered with row cover; then, as winter progresses and the crops need more protection, a layer of green house plastic can be placed over the top of the row cover and held down with landscape staples, sandbags, or rocks. Be aware that this tunnel will not be able to withstand consistent or heavy accumulations of snow and ice.

High tunnels, or hoop houses, are sturdy, permanent structures tall enough in which to walk upright, and large enough to plant several rows of crops. Constructed of polyethylene over a rigid pipe framework, high tunnels utilize passive solar heat and protect crops from damaging weather conditions, such as frost, wind, excessive rain, and snow. Crops are planted directly in the ground and can be cultivated in the high tunnel all year round. Roll-up or roll-down sides provide ventilation in the summer, and on some models, the ends of the tunnel can be rolled up throughout the warmer months. Because most high tunnels typically do not utilize electrical systems, such as heaters and fans, they are much less expensive to build than a conventional greenhouse. Our hoop houses allow me to grow flowers and vegetables two to three months earlier in the season. Their protection also produces taller, straighter, and undamaged blooms that I can sell at a premium. Even though the structures are unheated, the soil stays a good 10° to 15°F warmer than the soil in my field, making it possible for me to overwinter tender perennials that would perish outside. High tunnels can be constructed in many sizes. Several companies sell kits that contain all the parts and materials needed to construct a high tunnel (we purchased our kits from FarmTek and have been very happy with the results). Be sure to investigate opportunities for grants or other state agricultural monies that could pay for your new high tunnel.

Cold frames are a wonderful way to take advantage of passive solar heat without having to go to great expense or trouble. A cold frame is basically a box with a glass or plastic top. Once again, you can make your cold frames fancy and high-tech—there are many books and Internet sites devoted to cold frame construction—or you can do something as simple as surrounding a garden space with straw bales and covering the top with an old window. Most cold frames are somewhere in the middle, a base structure made from boards, often with the back side several inches higher than the front to allow the sun to angle into the box (the cold frame should face south to capture the most sunlight), and a clear hinged lid made of glass or plastic. Cold frames sit directly on the ground and you will want to enrich the soil with organic matter just as you do your garden beds. I love my cold frames because they allow me to eat my own fresh spinach all winter long. I grow lettuces, herbs, and other greens in them in the spring, but even if I only grew the spinach, they would be worth having around. Of course you can't grow just anything in the extreme cold of winter (sorry, no tomatoes!), but there is a nice selection of cold-tolerant crops with which to experiment (many are in the Brassicaceae family—yum!—arugula, kale, broccoli raab). You can also use your cold frame to start seedlings, or as a place to move your newly transplanted seedlings so you have more room inside. Be sure to provide adequate ventilation—it gets very warm very quickly in a glass box!—I simply prop my lid open with variously sized bricks, and on mild days, open it up all the way. If you are expecting an especially cold night or two, you can fill plastic jugs or water bottles with hot water and place them at intervals throughout your cold frame. The heat in the water will dissipate slowly and help keep your plants from freezing. You can also cover your cold frame with an old blanket.

Conventional Rows

Farming conventionally means the frequent tilling and working of the soil using machinery. This is generally reserved for 4-acre or more plots of land. The usual process is to first plow, using a tractor. You wait a week and then go back and disk the rough ground (you've probably seen this tractor attachment with numerous circular metal disks in a row), which levels it out. You might have a tiller attachment (just a tractor version of the walk-behind tiller) for your tractor and you till up rows. Or you use a walk-behind tiller to create the rows. You generally plant in mostly level ground. Then when the season is over, you plow under the crops and plant cover crops. The next spring, you plow under the cover crop and start over.

DRIP IRRIGATION

Drip irrigation is your best option for watering your plants. Any garden of size on a farm will take too many hours to hand water. And overhead watering with sprinklers invites fungal diseases and wastes a lot of water through evaporation. Drip irrigation is cheap, easy to install and replace, and puts water directly where it's most needed, right into the ground.

1.
You start with a pressure regulator and filter assembly that hook directly to your faucet. You can buy this as a set.

2.
From the regulator runs your main lines and they can be configured with elbows to run across the top of your rows. The drip tape attaches to the main line and runs down your rows.

5.
An elbow connector for the main line

6.
A special tool drills out a hole in the main line where you want your drip tape to run out from.

3.
It's easy to cut the main line where you need it and attach screw-on connectors and elbows.

4.
The screw-on connectors can be attached by hand.

7.
Attach a valve to the main line. These also screw in by hand.

8.
Attach the drip tap to the valve and run the drip tape down your row. Generally, two rows of drip tape is enough for any raised bed.

9.
Close off the end of the drip tape by cutting about an inch off of it. Then fold the drip tape over itself two or three times and slide the one inch pice over it.

Pest Control

Become familiar with the various beneficial bugs that inhabit your garden. They are your allies in controlling the populations of plant-eating bugs. The population of insects in your garden is complex and interconnected. Large-scale interference, by attempting to eradicate pests, upsets the balance of this network, and we cannot know the extent of unintended consequences.

Avoid using any pesticides or herbicides. Remember that any chemical pesticide or herbicide, even if it is organically certified, will take its toll on the beneficial insects as well as those that are causing you problems.

If you see a bug you can't identify, look it up on the Internet. If you are able to take a picture, you can send it in to your local agricultural extension agency for their entomologist to identify. It is all too common for someone to see a startlingly marked or fearsome-looking bug and react by squashing it. The prehistoric-looking assassin "wheel" bug is one such beneficial predator with a scary appearance (it does bite if threatened, so approach with care) that eats many harmful bugs in your garden.

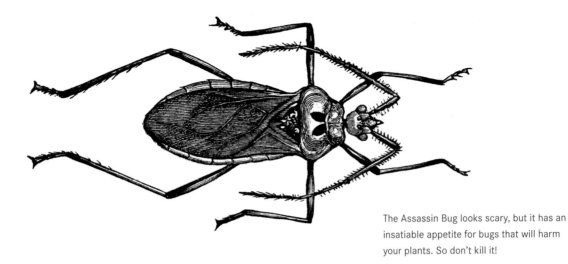

The Assassin Bug looks scary, but it has an insatiable appetite for bugs that will harm your plants. So don't kill it!

STRATEGIES FOR DEALING WITH DESTRUCTIVE BUGS

- Plant enough of a crop that you can afford to lose some of it to bugs. Even beneficial bugs can cause plant damage. For example, the caterpillar of the Monarch butterfly loves to feed on asclepias—I grow a couple of different varieties of the plant as specialty cuts for the bright flowers and interesting seed pods.
- Practice crop rotation.
- Raise healthy plants. Healthy plants are more likely to recover from a bug infestation and less likely to succumb to diseases that bugs can spread.
- Hand cull the bad guys. Scoop bugs into containers of water—this is my preferred method for Japanese beetles. I then feed them to my chickens. Soft-bodied insects, such as cucumber beetles, are easy to squash as you're working in your garden.
- Keep free-range chickens.
- Protect vulnerable crops with row covers. Made of material that lets in water and sunlight, they are light enough to drape over young crops, or suspend over wire hoops. Row covers can be purchased online through seed companies.

FARM ANIMALS

CHAPTER EIGHT

Dozens of books exist for each type of farm animal you might be interested in raising. And there are different considerations for each. For instance, caring for a milk cow is different from caring for a meat cow. I cannot give you in a single chapter all the complete guidance necessary to raise all types of animals. But what I hope to provide are the key considerations that will help you determine which animals might be right for you and your farm. And more important, I hope to share the joy that animals can bring to your farm and your life. If you want to get down into the weeds about specific animals, so to speak, I've provided a list of recommended reading for you on page 331.

When we first moved to our farm, we would sit on our front porch, admiring the space and view of the 8-acre pasture that stretched out in front of us. But after we became acclimated to having space, we realized that something was missing from that pasture. Somehow, it did not feel complete. A pasture needs grazing animals as much as the animals need the pasture. A pasture left on its own will return to its original form—a forest. This is not a bad thing. But there's a certain beauty inherent in wide-open spaces, and besides, we already had a good amount of forest. We soon realized that a farm ecosystem needs animals to stay healthy. That's when we went out and purchased our two donkeys from a farmer up the road.

Caring for animals is communal. You do not just commune with the animals and the land. You will find that you need the help of other people, too. And so animals will help you commune with like-minded animal lovers and your neighbors (if they're

animal people, of course). So, if you have a farm, you should keep animals, even if they are only pets. Yes, you can make a living by raising animals. But there are many other rewards awaiting you just by sharing your farm with them.

Animal Farmers

Women farmers have a special relationship with their animals. "Taking care of another living creature is very similar to raising kids," says Susan Wise-Bauer, who currently raises sheep at Peace Hill Farm, and has farm schooled her four children. "You get the same rewards, knowing that they are fed and happy." She goes on to describe how there's a visceral feeling she gets when she lets her sheep out into a fresh paddock of clover. It's similar to the feeling you get when you let your children indulge in dessert and you take satisfaction in their pure pleasure.

On a larger scale, Temple Grandin, in all her years observing animal processing facilities and researching problems with animal cruelty at those facilities, reports that women were never the ones directly harming the animals. They may have been a secondary offender, but they were never the ones creating the problems. She attends many "Big Ag" and "Small Ag" meetings across the country and has observed that female animal farmers tend to raise smaller animals, such as goats, sheep and chickens. Sure, there are plenty of women raising cattle, but the real rise in women animal farmers is on the small-farm end and includes mostly small animals.

It's easy to imagine the female desire to nurture would translate well into animal husbandry (is it time to retire this term?). All the women I talked to echo these sentiments. Women tend to treat animals gently and have a bit more patience with them. This isn't to say that there aren't men who share these

qualities—and I know quite a few that do. And perhaps a big factor in treatment of animals is the size of the farm and not necessarily the gender of the farmer. A large farm with thousands of animals would certainly make it easier to forget that those animals are individual living beings and not just numbered "stock." And as we've observed that women are leading the charge into small and sustainable farming, it makes sense that they are the ones developing more humane methods to raise animals. It's not hard to imagine why it took Temple Grandin, a woman, to develop livestock handling systems that took into account how animals see and feel the world.

THE GREAT DONKEY ESCAPE

A gloomy morning in mid-November

My morning chores are routine and don't change much throughout the season. Although December approaches, I haven't yet started feeding the llamas and donkeys their winter food supplement of a flake of hay every morning. I begin with the house/yard animals, currently five cats and one dog. They get fed first thing so they'll stop pestering me, then on to the chickens. We have two separate small flocks and I open the coop for each and give them a portion of feed, check their water, and take a head count to make sure everyone is still around.

We've had quite a bit of rain lately, so the yard is a marshy expanse of squooshy mud and dormant grass. I slog over to the pasture gate and see the llamas off in the far field across the pond. They're chewing grass and keeping an eye out for field interlopers, such as foxes or stray dogs; I don't really think much about not seeing the donkeys; they could be in their shed, or hanging out in the wooded glen by the creek. It's not terribly cold out, but a misty rain has begun to fall, and I'm quickly getting damp in my fleece jacket and no hat.

Inside feels cozy warm and I pour some coffee and peruse my to-do list. My neighbor's car grumbles into life, and I listen to her back out onto the road, hoping her little Saturn wagon will make it through another year. It's then that I realize what I don't hear. There is no screeching bray from our donkey Whipper. Any car coming or going up our small dirt road receives this message of warning—"Hey, I'm here, and I'm loud." Whipper often brays for no reason at all, but he always brays at the noise of a car on his road.

Somewhat concerned, I put on my jacket and a hat, and head out to the field. "Heeeeeeere, Donkey! Heeeeeere, Donkey!" The llamas prick their ears and start walking over, but no sign of either donkey. I head over to the shed—no donkeys. I start toward the wooded glen, following the fence line of our pasture. We've got almost 10 acres fenced in with three rows of high-tension wire. Donkeys and llamas aren't particularly interested in finding ways to escape, so I'm surprised, but still not in all out panic mode, when I find a section of the fence that has had its top two wires snapped by a fallen tree. The multitude of hoof prints overlaying one another on both sides of the fence quickly propels me into a state of heart-pounding anxiety. I do a jog around the fence perimeter in case there is another breach, and for the slim-to-nothing chance that the donkeys did not decide to go on an adventure and are still somewhere in the field.

They are nowhere to be found. I know I should repair the fence, right now, so that the llamas don't decide to high-tail it as well. In the shed I find I don't have the wire and the wire clips that I need and decide to make do with a roll of nylon cord. It will have to do. Wrapping the cord tightly around trees and posts to create two rows of temporary fencing seems to take an eternity, but I know it's the right thing to do. I then hurry back to the house to begin the phone calls.

The first call is to Michael, who is working in New York City this week. The second call is to the County Animal Control to report that the donkeys are on the loose and find out whether anyone has spotted them. The animal control officers don't have any information for me, but say they will call if they find out anything. They take down my information.

In the meantime, Michael has called our friend Mark who lives a few miles down the road. Mark arrives and we set off together to search the 10 acres of our woods in an attempt to find out which direction the donkeys have gone. The donkeys are large animals, and the ground is wet, so for a while we track them easily along the trail in our woods. They seem to have had a little scuffle near the abandoned truck next to the pine grove, and then headed off into the woods.

Now when I say scuffle, I should provide a little background. Our donkeys are brothers and neither of them has been neutered. They haven't seen a jenny since they were a couple of years old. Most donkeys are known to play a domination game where one mounts the other and bites the back of its neck. The donkey being mounted

bucks until the other donkey falls off and they have some face-to-face wrestling on their hind legs. It's all perfectly natural, and even more so for donkeys that don't have a jenny to mount in the normal course of their lives. Inevitably, the donkeys like to play the mounting game whenever someone they don't know, especially small children, are around. They must have been feeling a little playful after their big escape.

We go a bit farther, but the ground becomes impassable as the network of small creeks have overrun their beds, flooding this part of the woods and making it impossible to see any tracks. The good news is that this is the side of the property that is farthest from the busy, 55-mile-an-hour traffic of Route 22. We head back around to the front of the property just in time to run into Mike and Mary, who have also been alerted by Michael and rushed over to aid in the search. Mary offers the use of a trailer that she and her husband, Harold, are trying to sell. Mike offers to hook it up to his truck and help me load up the donkeys when we find them.

I return to the house, check for messages (no cell coverage, so that's the only way to know if anyone called)—none—and grab my car keys. I drive down all of the roads that are within 3 miles of our property on this side of Route 22. It is a weekday and not many people are out and about, but I ask the few people I see whether they've spotted two gray and white donkeys, and give out my phone number. I return home to check for messages. Still none.

When the phone finally rings, it's Animal Control:

"Uh, Ms. Lev-a-tin—." He stumbles over the name and I cut him off before he finishes.

"Yes, this is she."

"You called in this morning to report two missing donkeys?"

"That's right."

"Someone here took a call early this morning from a woman who reported that there was a pair of white horses . . . ." He pauses to call out to someone on his end of the line, ". . . what did she say?"

In the background I hear a woman's voice, "They were fornicating in her yard."

"Uh, ma'am, the caller said they were . . . procreating in her yard."

I suppress a twinge of embarrassment that prompts me to apologize for my animals, and say, "Yes, that sounds like my donkeys."

"Well, this was over near Longwood Drive, so we'll send someone out there to look around, and we'll call you with any more news."

I thank him and hang up.

Longwood Drive is about a half-mile away off of Route 22, but it is accessible from the back of our property if one traipses through the woods on a diagonal. From where we tracked the donkeys' prints this morning, it makes sense that they headed in that direction. I hope they are still over in that area, and not "procreating" in anyone else's yard. The day continues to be gray and drizzly as it heads into early afternoon. The next call I get is from Bradley McGhee, the chief animal control officer. Tyler and Whipper have been found. They are in the front yard of a home off Longwood Drive and Mr. McGhee will be at my house in a few minutes to lead the way over there.

I call Mary to let her know the donkeys have been found and that I'll need the trailer, and Mike's help, soon. Relief has made me lightheaded. I grab my keys and my cell phone, on the off chance that I can get a signal a few miles down the road, and head out the door. Scooping up halters and horse treats, I hop in the car to follow Mr. McGhee.

We turn into a cul-de-sac off Longwood Drive. There are six houses, each nestled in a spacious lot and tastefully landscaped. Outside of the last house on the right are my donkeys. Thankfully they aren't currently making a spectacle of themselves; instead, they are happily munching away at the bark of the trees that border the property. Bradley McGhee is a tall man in his late 40s/ early 50s. I step out of my car, expecting a bit of a lecture about keeping my animals fenced in, or at the very least, a stern reprimand for the time wasted and trouble caused by Tyler and Whipper and my inadequate animal care. But this is just another day at work for Mr. McGhee. As he helps me secure the donkeys, each to its own separate tree, he chats about how his grandfather had a donkey. His calm manner helps to settle my nerves. He's been working for more than 12 hours already, and once I let him know that I've got friends coming to help me get the boys home, he heads off for some much deserved rest.

I spend several minutes talking with Tyler and Whipper, giving them each a couple of treats, and looking them over for any signs of travel damage. Being donkeys, they show no indications of remorse for their behavior. Because they never knew they were lost, they don't even appreciate being found. I wonder what their adventure was like, how it felt for them to be out on their own all night, wandering through woods, across creeks, and among houses. Overall, the sense I get from them is that they expected me to show up eventually.

I return to the car and try my cell phone. No signal. I look around the cul-de-sac. Most of the houses seem closed up tight, warding off the chilly, gray day (or barricaded against a donkey attack). One house across the way has a light on in the foyer, so I head over there and knock on the door. When the door is answered, I ask the woman whether I can use her phone, indicating the donkeys and explaining my lack of cell phone signal. She smiles kindly and says she'd watched as we tied the donkeys to the trees and asks me what happened. I give her the short version and then call Mary's house. I give Mike the address and directions, and he says he'll be there in about 20 minutes. Thanking the woman, I return to my car to shelter from the drizzle.

Mike arrives with the horse trailer in tow, and this is where the real trouble begins. As a younger man, Mike spent about 15 years working with race horses. He doesn't like to put up with any shenanigans from large animals. Donkeys don't really care about anyone's preferences. The stubborn in the phrase stubborn as a mule comes from the donkey side of the mule's genetics. My donkeys have never been trained to go into a trailer, so they instinctively resist being led into an unfamiliar, small, enclosed space. We give Tyler the old pull-and-push technique with no results. After about 15 minutes of this, Mike's frustration mounts and he decides we should try to put Whipper in the trailer first.

Whipper is larger than Tyler and we have the same trouble getting him into the trailer. I can tell that Mike is angry, and I appreciate that he is trying to control his temper. I feel badly because I am just not strong enough to be really useful, but I'm not sure that strength is actually what is needed here. I tell Mike to let me try to get Tyler in the trailer first. Tyler and Whipper are half brothers, with Tyler being older by two years. Wherever Tyler goes, Whipper will usually follow.

I take Tyler's lead rope and walk him around

in a slow circle a couple of times and then lead him up to the trailer ramp. He balks, so I walk him around in another slow circle and try again. This time he hesitates but I don't pause; I keep pulling the lead into the trailer. He's up the ramp, and Mike pushes him in the rest of the way and we secure him to the back.

By now we've been out in a steady cold drizzle for almost an hour. Our hope is that Whipper will go into the trailer willingly now that Tyler is there. Of course, this doesn't happen. And so the battle begins between Mike and Whipper. On the positive side, I am there, so Mike can't get too heavy-handed with Whipper, and donkeys are not naturally aggressive—so the likelihood of any of us getting injured is small. The next 20 minutes or so pass in a discouraging mélange of pushing, heaving, threats, and cursing. Mike and I are both getting cold, we are tired, and we are getting nowhere with this donkey situation. Mike tries calling a couple of our friends to see if anyone is available to help us. No luck. His last call is to our friend Blue, and not only is he home, he comes right over to help. At first glance Blue might not seem like the optimal choice for maneuvering donkeys. Slight of frame and in his 80s, an outsider would wonder what use he could be in this situation.

Blue saves the day. He is calm. He is cheerful. He retains his sense of humor. And eventually, after the same tactics we'd been using all afternoon, Whipper submits and stumbles into the trailer. Did Whipper simply get tired and give in? Maybe. But because Mike and I had Blue there to keep us sane, we didn't give in, and we got those stubborn donkeys home.

Michael returns home from New York City and is delighted by the whole "donkey rescue story." Over several beers at Mike's place, Michael gets to hear the story from both Mike and Blue, a couple of times. I'm so grateful for the help. Mike certainly didn't have to help me, and of the people I know in this neighborhood, Mike is by far one of the busiest and hardest working, with the least free time to spend wrangling donkeys. But he did help me, and I never doubted that he would. However, I wouldn't blame him if he's reluctant to pick up the phone next time he sees my number on a cold and drizzly day when Michael is out of town.

WOMAN-POWERED FARM

Animal Partners

Animals are an almost indispensible partner on your farm. Any type of farm animal you can think of can contribute something special that makes a farm more sustainable and enjoyable. Cows, cats, sheep, chickens, dogs, pigs, llamas, alpacas, goats, turkeys, ducks, and horses, all have unique natural abilities that can help you farm and bring joy into your life. Animals provide food, soil fertility, protection, and companionship.

Your Animals, Your Dinner

The most sustainable and easy source of animal food would be the eggs of your chickens. With enough room to roam, chickens can pretty much feed themselves except in climates where snow completely covers the ground for part of the year or in a desert. Providing some quality feed for them as a supplement will help them produce quality eggs without much interruption for most of the year. And other than providing them with shelter, water, and a bit of supplemental food, you can create an environment for your chickens and yourself that is virtually maintenance free. Chickens, along with bees, are the best option for urban farms, for the obvious reason that they can be kept in small spaces.

You can, of course, eat your chickens. Chickens are easy to put down, clean, and store, if you know what you're doing. And you should have no trouble finding local farmers who would show you the proper techniques, as long as you are willing to put in some time helping them out on a processing day. You can also eat cows, sheep, goats, pigs, and so on. What you should consider, though, is how will you have those animals put down and processed? It's a much bigger commitment than for a bird. If you plan to sell the meat you don't eat, you'll need a certified processing plant that can require transporting your animals up to hundreds of miles. Once you feed and raise them and then ship them off to be processed, are the costs and efforts worth the benefit?

Cows can eat over 100 pounds of feed and drink 30 to 50 gallons of water a day. Depending on how long you keep your cows, it can take 15 pounds of feed to produce 1 pound of beef. Sometimes it takes more, sometimes less. Along with the gaseous emissions they release into the atmosphere, it's a well-known fact that cows are a negative drag on the environment because the meat they produce requires so much more in the way of inputs. But cows can also be a valuable piece of a fully sustainable farm system that includes a rotation of other animals, grasses, and crops.

More than Just a Good Meal

Surprisingly or not, food is not the best long-term contribution that an animal can provide. Animals are the most efficient converters of the sun's energy into useable nutrients. It's hard to create healthy soil on a farm, at least at any useful scale, without the aid of animals. Plants use the sun and water to grow. Grazing animals then eat the plants and convert them to useable nutrients in the form of manure that is well balanced and full of minerals. This new energy is then spread on our fields by the animals themselves, improving the fertility of the soil, or we gather it up and use it in our gardens. If we were to attempt this on our own, we would need a large-scale composting operation with large amounts of food and plant material. And the best compost you can get incorporates animal manure, so you'd need to import manure, if you don't have animals. Why not just have your own fertilizer machine—a grazing farm animal?

People ask me all the time, "Why do you have donkeys and llamas? What do you do with them?" The answer is that we coexist. We enjoy each other's company. I keep them safe from the dangers of the human world (they're quite capable of handling the natural world) and they provide me with the black gold that makes the vegetables and flowers grow—their manure. They provide food and income in a not (at first) obvious way that most people think of farm animals.

The manure of llamas, or "llamanure" as we like to call it (sometimes with a French accent, just for fun), is particularly useful. Because they are ruminants, the manure is fully processed and cold when it comes out in the jellybean-like pellets. It's called cold manure because it's ready to be put directly into the soil and useful immediately for your plants. Hot manure, which comes fresh out of most other grazing animals, such as our donkeys, is full of hot nitrogen that will burn your plants if you don't let it age or compost it first. We actually bag up the

extra llamanure we don't use and sell it at the farmers' market.
The money we make covers the cost of the supplemental feed
we give the llamas in the coldest part of winter. The donkey
manure needs to be aged and composted, but because they are
larger, they provide us with more manure volume. And one very
helpful talent that both the llamas and donkeys have, which we
only discovered after we bought them, is that they like to poop
in the same spots over and over again, creating nice piles of easily findable and scoopable black gold. They're probably baffled
by our fascination with their excrement, but we're thankful they
have a good enough sense of humor to pile it up for us to find.

Your animals can also serve as an extension of good fencing, the best line of defense against predators. Dogs, llamas,
and donkeys all serve as livestock guardians. Our llamas and
donkeys regularly chase off roaming dogs. We've also seen
them chase foxes and deer out of our field. The best guardian
is a good dog. And a good dog will deter unwanted humans as
well. As a woman living in the country and at times alone while
my partner travels, I find the security of a good strong dog to be
invaluable. Casey, our golden retriever mix, alerts me whenever
anyone comes to the front gate and also patrols around the
chicken coops at night. So far, we've had very few problems with
predators attacking our chickens, or me, for that matter.

Keeping Animals

Where do you begin if you want to keep animals on your farm?
As with all the other aspects of farming, I'd suggest starting
small. When we first moved to the farm, the only animal we
brought with us was a cat. We learned a sad lesson about the
adaptability of a city cat to a rural setting. Our cat was used to
being inside and neighborhood fences kept out dogs. When we

came to the farm, our cat was not used to being in such a wild
place. She was killed by a roaming dog, which taught us the lesson that farm cats need to be raised on the farm to develop the
proper skills to stay alive. This is true for other animals as well.

We learned another lesson about predators very quickly.
The previous owners had left behind a rooster and hen. They
were a happy couple that free ranged in the yard. I named them
Ted and Bev. They had no coop, as the previous owner had a
couple of dogs that slept outside at night to keep away predators. Bev was able to fly up into our holly tree and roost for the
night. Ted was too heavy to fly, so he slept on the front porch.
Every morning we awoke to the sound of Ted's garbled crow
outside our bedroom window. Until one morning we didn't. We
only found a trail of white feathers through the field and into the

woods. I was so guilty about not providing for Ted's safety that I
named our farm after him—Ted's Last Stand (see page 298).

I don't tell these stories to discourage you from keeping
animals. They are only cautionary tales to keep you from making
the same mistakes we made. Animals are one of the true joys
of living in the country. But do not romanticize them and make
sure you do your homework before you jump in. The best animal
to start with is a dog. You need a guard animal before you start
adding others. Then maybe a cat or two, which you'll want to
keep outside most of the time to keep rodents away from your
house. Once you're comfortable with your domestic animals,
you can move on to farm animals, and the best place to begin

is with chickens. They are easy to care for, inexpensive, and provide you with delicious eggs. Once you're comfortable raising some chickens and you've dealt with their housing, protection, feeding, and inevitable deaths, then you might be ready to take on larger, grazing animals. What you choose from here is your own personal choice and all depends on what you want to gain from the animals you keep.

There are many wonderful books written about raising every kind of animal you will find on a farm. For several suggestions, see Resources (page 331). The first book that you should read is Temple Grandin's *Animals in Translation*. Dr. Grandin's book about how animals see and experience the world is an essential primer for anyone who wants to raise animals.

You must first decide whether you are keeping your animals as pets and permanent members of your farm ecosystem or you want to become an animal farmer. You can, of course, practice a combination of the two, but there are very different considerations in the care of pets versus livestock. This chapter will provide you the basics of care for a variety of animals and the considerations you may want to make in deciding whether you will raise them as livestock. You will not learn everything about raising any of the animals discussed and I won't touch on every possible animal you could keep on a farm. I hope to provide you with a primer to the most common farm animals, how they might fit into a farm system, and what issues you might need to consider while deciding whether raising a particular animal is right for you and your farm.

The absolute best advice I (and every animal farmer I know) can give you if you want to raise animals for a living is to suggest that you work for at least one season as an intern or an employee on another animal farm. There's only so much you can learn about the care of living things from books. Hands-on experience is the key. Not only will you experience the full seasonal cycle of caring for animals, but you'll quickly learn what you *don't* want to do. After waking at dawn every day for a season in all types of weather to milk goats, you may decide to raise an animal that's less labor intensive. After being attacked by a sow protecting her piglet, you may decide that risk of bodily harm is not worth the bacon. After finding a chicken massacre in a coop because you didn't secure the fence properly, you may decide that such helpless animals aren't your cup of tea.

Nothing replaces the value of experience. And you can save yourself a lot of time, money, and heartbreak by working on another farm to find your calling. If this isn't a practical route for you to take, start very small and find yourself some good mentors in your local farm community that you can rely on for help and information.

It's common sense that food, water, and shelter are the basic needs of animals on a farm. They are the basic needs of humans as well. Following are some other rules of thumb for all animals that will help keep your animals safe, healthy, and happy.

Prepare for the Entire Life Cycle

Before you bring animals onto your farm, you should be prepared for the entire life of that animal. If you are keeping animals as pets, to ride, or for any other reason except for food, you need to consider how long they are going to live. Horses and donkeys can live to 30 years or more. Llamas in our area generally live to be almost 20 years old, and sometimes older. Even our original chickens, Ted and Bev, lived to the ripe old age of 11. Plan accordingly and make sure you have plans to care for them through their entire life cycle, if that is your goal. You can always sell or find good homes for animals. But you'd be surprised how hard that can be, especially for large animals that take resources to feed, such as horses.

And if you plan to raise animals for food, you should first find out where you will process them. Chickens can be pro-

cessed right on the farm. But larger animals need to be processed in a USDA-certified facility if you plan to sell the meat.
And processing plants have seen the same struggles as small
farms in the last few decades. Consolidation and fewer small
farm customers have put many out of business. You might have
to trailer your animals to facilities that are hours away. So,
before you buy any animals find out where the facility is and find
out how you will get the animals there and back. This can add
considerable cost to processing animals for meat and sometimes eats up any profit you may have hoped to gain.

SIZE

As we see anecdotally, many women getting into farming are
choosing to raise the smaller species of farm animals—sheep,
goats, alpacas, and chickens. You should consider what size animal you are comfortable handling. A sheep or goat, for instance,

can be picked up and even transported in the back of a car, if necessary. And being timid around large animals is a recipe for injury. But just as important as the size of the animal is the work that it takes to take care of the animal. It takes a large volume of feed and water to feed large animals. So, you will be carrying and transferring more weight when you care for them. On a daily basis, this can become difficult for anyone. One pulled muscle or slipped disk in your back and you could have a hard time keeping up with the care of large animals. That's not to say any woman can't care for any sized animal. It's just a consideration each woman needs to make for herself.

FENCING

Proper fencing is as important to animals as food and water. There are certain types of fencing that work best for each type of animal. Make sure you match the fencing to the animals you are keeping, to prevent escape and/or injury. Never use barbed-wire fencing for any animal, except cattle. And there are other alternatives for cattle, too.

VETERINARIAN/FIRST AID

As soon as you decide what type of animal you will keep, find a local veterinarian who specializes in that animal. It's best to ask around and go by word of mouth. That is, except for chickens, for which there's really no reliable or cost-effective veterinary help that I've ever found. And perhaps as another sign of the changing tide in agriculture, I've found that every vet we've ever had has been a woman—from our small animal vet who does house calls for our dogs and cats, to the donkey vet who came and treated a growth on Tyler's butt, to the three llama vets who have visited our farm over the years and also specialize in cattle. Maybe it's just coincidence, but anecdotally, I hear that young women are dominating all the entry-level veterinary jobs in our area and are slowly taking over practices from the older, graying men who used to be the local animal docs.

Just as important as finding a new vet is learning to do as many basic veterinary tasks yourself as possible. Learn to give shots (for deworming and tetanus), dress wounds, and even sew up small wounds with a stitch or two and some local anesthetic. Consider taking an animal first aid class at your local community college. Or ask another farmer or veterinarian to teach you. Even if you have to offer to pay or trade with them for the knowledge, it will pay off in the long run. You'll gain some valuable confidence working with your animals in this way and you'll also save yourself a lot of money; many times it will be the difference between making a profit on an animal or a loss.

Keep a first-aid kit (all these supplies are available at your local feed store) that includes:

- Iodine
- Antibiotic ointment
- Fly-repellent ointment
- Gauze wrap and bandages
- Vet wrap
- Latex gloves
- Digital thermometer
- Syringes and needles that fit the size of parasite doses particular to your animal
- Rubbing alcohol

- Wound ointment
- Digital animal thermometer
- Head lamp and penlight
- Local anesthetic
- Optional: stethoscope, and sewing needles and medical thread for sutures

SAFETY QUICK-RELEASE KNOT FOR TYING UP ANIMALS-

this is an essential knot to learn. You should always tie animals up with it. If they panic or fall down and you need to set them free, you can pull the end and the knot will release.

Some animals, like horses and donkeys, can be treated for parasites with a gel that you spirt into their mouths. Others need to be injected with a syringe.

PARASITES

Parasites are perhaps the biggest threat to the general health of all farm animals. Parasites are generally worms or insects that attach themselves to the outside or inside of an animal's body and feed off them. Ticks and fleas are examples of external parasites and worms are examples of internal parasites. External parasites can be easily dealt with once you find them with a topical treatment. You can find the ointment for various types of external parasites at your local feed or farm supply store.

Internal parasites are the most dangerous to animal health. All animals have some level of parasites and they aren't always dangerous. But they do cause stress to animals, and if you have

WOMAN-POWERED FARM

an already weak animal, parasites can speed up its demise. You should familiarize yourself with the parasites that are likely to affect the type of animal you are keeping and follow a veterinarian-recommended prevention and treatment.

We learned the hard way about parasites from one of our llamas. We became a bit lazy a couple of years ago in deworming Rico, our llama. We felt that the drug companies probably overprescribed in their recommendations. But one day we saw Rico having a hard time walking and losing his coordination. He had ingested the meningeal worm that is transferred through the feces of deer. It attacks the nervous system of llamas and can be fatal. Luckily we caught it in time and through an extensive dosing of deworming medication over a full week, he mostly recovered. But he now has permanent nerve damage and it was our fault. So, don't play around with parasites. Prevention is the key.

Some dewormers come in the form of a pill; some require a paste that you squirt into the animal's mouth; some can be put into an animal's food; and some, such as what we used for our llamas, require you to administer a shot. The frequency of dosage is generally every six weeks to six months. It just depends on the type of medication and animal. Figure out what is recommended for your animals and don't put off that chore for another day. And if you aren't confident giving a shot right away, have an initial visit from a vet and ask him or her to show you how.

Besides drugs, clean living conditions are the best way to prevent parasites. Parasites are also transmitted many times from wild animals. Keep your animals in clean, dry enclosures that exclude wildlife as much as possible and you'll keep parasites under control.

SOCIALIZATION WITH OTHER ANIMALS

Always have more than one animal of any type living together. All animals are social and they can quickly get depressed and their health will suffer if they're lonely. Animals can also become aggressive and skittish without a companion. Exceptions to this rule are animals that live in your house and get your attention much of the day, such as a cat or dog. Also, you typically want to keep bulls or unfixed males of any breed away from others. Depending on the animal, having two unfixed males housed together is generally a bad idea.

Some animals of different species seem to truly enjoy each other's company. Donkeys have always been used to calm horses. And many donkeys form bonds with animals they guard, such as goats. Dogs and horses don't always mix, but sometimes they become lasting companions. Our dogs, cats, and chickens all mingle together. They aren't best friends, but they tolerate one another and never get into any type of dangerous conflict. The best way to have animals that intermingle in this way is to introduce a new species of animals when they are very young. All animals of good temperament recognize babies of any species and generally treat them gently. As they grow up, they just become part of the pack.

When introducing animals to other animals on your farm, keep the new animals in an enclosure or pen adjacent to the other animals you're introducing them to for a couple of days. Animals need to get used to one another's smells. They also need to be able to get up close to the other animals through a fence so that they can satisfy their curiosity without fear of attack.

SOCIALIZATION WITH HUMANS

It's a good idea to have contact with your animals at least once a day to build trust and make it easier to handle them when you need to. It's also the only way to keep on top of any changes in their health. Pet them; don't pat them. As Temple Grandin points out in *Animals in Translation*, some animals interpret patting as hitting. Call them over with food. Bribery goes a long way. Give them treats occasionally and talk to them in a low, soothing voice. Compliments are appreciated.

Another technique I learned from Dr. Grandin is to put yourself down on the level of your animal. Once your animal is comfortable and calm around you, sit on the ground, or even lie on the ground. Most animals spend much of their day with their face next to the ground, grazing. When you lie down, you're eye to eye with them and they see you as a part of their regular world. You're no longer the tall creature looking down on them. This is also a great perspective for animal photography.

PASTURE

The more space you can give an animal of any kind to roam and graze freely, the better it will be for its health and the health of your farm. Rotational grazing is the key to improving the health of your soil, pastures, and animals. Overcrowding of animals stresses them, breeds disease, and forces you to spend more money on supplemental food. If you cannot provide pasture for your animals, then you should reconsider raising animals. The exceptions to this rule are animals that can make themselves obese, such as donkeys, and that might need to be kept in small paddocks to keep their weight from becoming unhealthy. Also, if you are keeping chickens in an urban environment or in a place with a predator problem, then enclosing chickens instead of allowing them to free range is the most ethical thing to do.

Before introducing an animal to a pasture, make sure to walk it in search of anything dangerous. Look for pieces of metal or nails it may step on, sharp pieces of fencing that it might get snagged on, deep holes it might step in, and any poisonous plants. Have your cooperative extension agent or another experienced farmer come out and help you identify plants that might be poisonous.

ROTATIONAL GRAZING

This form of raising animals on pasture hit the mainstream when Joel Salatin was profiled in the best-selling *The Omnivore's Dilemma*. Salatin, whose farm is just 100 miles or so from our farm, often calls himself a grass farmer. This is because he practices a form of rotational grazing that is most concerned with building healthy soils and grasses. The basic concept starts with moveable, solar or battery-powered electric fencing. The idea is that you confine such animals as cows to a certain area of field where they can eat down the grass and convert it to fresh manure in that area. When the grass is eaten down, you move the fencing and the animals onto fresh pasture. The spot they left behind is then rested, the fresh manure is slowly composted back into the soil, and the grass regenerates. To make this even more productive, you can rotate chickens into the area where the cows or larger animals have just grazed.

The chickens will scratch and spread the manure while eating the bugs in the manure. The chickens become living tillers and work the manure into the soil. When done, the pasture has been fully fertilized and aerated and you will see some of the healthiest grasses then grow in its place. You can rotate any animals, using the moveable fencing around your fields, and they will convert the grass to food for themselves while leaving behind food for the grass to regenerate stronger than it was before, even if you aren't rotating multiple species, such as cows and chickens. And the money you can save on supplemental feed makes this the most efficient way of raising animals. But you need a proper amount of space to make it work.

SUNSHINE AND FRESH AIR

These are keys to animal health. Especially in the winter, if you can provide opportunities for animals to soak up some rays and vitamin D, they will stay healthier. Obviously, too much sun can be a problem. So, shade in the heat of the summer (and in extreme cases, fans to circulate air) is as important as direct sun in the winter. This shade can be provided by fencing in a wooded area or providing a simple roofed or lean-to structure.

PROPER ENCLOSURES

There are recommended enclosures for every type of animal. But the type you choose depends on your climate and how you are raising them. If you are using rotational grazing then your housing will be small and easy to move. But a permanent chicken coop might be quite elaborate. Wet enclosures breed disease and parasites. Always give your animals a place to get out of damp conditions. Most animals are susceptible to foot or hoof disease in damp conditions.

You will need an isolation pen of sorts. There are always times when an animal is sick or you need to treat it for parasites. You should have a smaller pen that is easier to work with the animals in. This is also useful when you first introduce an animal to your farm. Keeping them in a smaller enclosure helps them to feel more secure and gives them time to acclimate. For our larger animals, such as the llamas, we use our dog pen (without the dogs in it, of course).

LEARN TO WRANGLE

You don't need a rope and a lasso to wrangle most animals. But it can certainly be a chore. Socializing your animals is the best way to ease your job. Come up with a call for them that you use every time and always have treats waiting for them. For our donkeys, we yell, "Heeeeee-Donk! Heeeeee-Donk!" For the llamas, we yell, "Heeee-llama! Heeee-llama!" The previous owner of our donkeys had named the mother jenny Willow. And she named her son (now our donkey) Whipper. She'd call them in at night by yelling, "Whipper Will! Whipper Will!" If you do this on a regular basis and earn their trust, you will be rewarded when you most need to wrangle them.

Our llamas can be hard to wrangle. When you have an animal that is skittish, you need to corral it into smaller and smaller spaces until you can get close enough to handle it. Sometimes, the llamas are in their shed and we just sneak up on them and block the entrance to the shed. To an outsider, we must look pretty funny creeping through the high grass, staying low and sprinting those last few feet before they realize we're coming. When they're out in the field, it can be more challenging. The dog pen is our goal, because it is small enough that we can corner them and get a harness on them. In preparation, we pull the truck across the side yard to the left of the house and pile lawn chairs across the side yard on the right. This after the first time we ended up chasing them around our house in circles. We then open the gate to our yard and corral them through it like cowboys without horses. Once in the yard, we close the gate behind them. One of us goes into the pen with treats and the other tries to corral them in. Sometimes we have to pile more chairs in areas they want to escape into or even string a temporary rope fence between trees. But eventually they will go into the pen on their own.

Chickens are hard to catch, but not if they are in an enclosure. You can corral them into a corner pretty easily with just your hands or a stick. You can buy a chicken hook, which is a long stick with a small hook on the end. The goal is to hook one of their legs. If they are in a big enclosure that makes it impossible to get close, just wait until dark when they've roosted for the night. They are completely docile and mostly blind and you can easily walk up and grab them. Other animals, such as sheep and goats, can be corralled by piling food in an area and then using the moveable fencing to surround them while they are eating. Or you might fashion a snare out of rope and lay it in the grass next to a pile of food. Once they are standing in it, pull hard and you might catch them by the legs. There's no one way to do it and sometimes you'll feel like Wile E. Coyote trying to catch the Road Runner. A herding dog is a wonderful thing. But ingenuity and treats are your best bet!

FOOT AND HOOF HEALTH

Get to know the feet of your animals. If an animal cannot walk, it cannot eat and it is in danger of dying. Horses are susceptible to founder and need regular visits from a farrier and shoeing if you ride them. Llamas and alpacas have toenails that need trimming every six weeks or they curl uncomfortably. Cows can get foot rot from wet and dirty enclosures. Sheep and goats both need hoof trimming to avoid disease. Dry, rocky soils are best for animals' feet and will keep you from having to trim or treat them so often. And wet and mucky soil is not good for any animals' feet or hooves, unless they are water buffalo.

FEED

it goes without saying that the better the quality of the food you give your animals, the healthier they will be. But you should also consider that if you are planning to eat your animals or their eggs, whatever you feed them will eventually become food for you as well. If you can't imagine eating the ingredients in your animals' food, then don't feed it to them. Read the labels of your feed carefully and stick to organic, if you can.

BREEDING

Until you've had a few years under your belt taking care of animals, you should stay away from breeding. If you are going to raise animals for meat, it's very easy to obtain young animals in the spring from a breeder, raise them through the summer and then process them in autumn. The most heartbreaking animal stories we've heard or experienced happen when a pregnancy or birth goes wrong. It takes experience to know what to do when a problem arises. Also, breeding is dangerous in some cases. There's nothing more dangerous in all of farming than coming between a pig sow and her piglet. Many farmers have lost their limbs or life in this situation. Eventually though, if you are trying to make a living, you'll need to breed your own animals as that is the most profitable way to raise and sell them. Again, I recommend an internship or apprenticeship on another animal farm before you begin breeding your own animals.

TRANSPORTATION

You should either have a trailer that can transport your animals or you should know someone from whom you can borrow a trailer. We don't have a trailer. The only time we've ever needed one was when our donkeys escaped the fence and we found them about a quarter-mile away (see page 225). We live in an area where we know lots of people with animals, so we're able to borrow a trailer if needed. If you are going to raise animals that you will eventually sell or process, then you should have a trailer. It will pay for itself very quickly.

LEARN TO LET GO

Gail Hobbs-Page, a local goat and cheese farmer in our area, says one of the most important things she had to learn to do was, "To let go." At first, she wanted to save every animal. She did not want to cull sickly animals or put sick animals down too soon. "You have to be able to send your babies off to slaughter. You have to be able to shoot things," she advises. So, think carefully ahead of time about how you will handle this part of the farm animal life cycle. Sure, you may just have farm pets and they can be handled by a vet and the few times you need to do it won't be cost prohibitive. But if you are going to be an animal farmer, you need to learn how to properly put your animals down, if needed, and you need to learn how to cull and send your animals off to slaughter. The only way to learn this is with the help of another animal farmer. Find a mentor or work/intern on another farm where you respect the treatment of the animals.

ERICA HELLEN AND HER INTERN ZAN MCKENNA

Free Union Grass Farm
Free Union, Virginia

Raising and selling chickens, beef, ducks, and pork is a lot of work for a couple. Erica and Joel Hellen did it without any help for the first few years, and now admit, "It's kind of miraculous that we didn't have any help. We could have burned out." They recently decided to hire interns, and now work only six days a week, with the occasional trip out of town.

The current interns at the farm are Zan McKenna and her fiancé, Dave Plummer. They are planning to start their own animal and vegetable farm on a piece of land owned by Dave's parents, and they are treating their six-month internship at Free Union Grass Farm as a kind of graduate farming school. By choosing a live-in internship, they are learning the trade they want to pursue, but not taking on any of the college debt that formal education can bring with it.

Like Zan, Erica's first experience with farming was as an intern. She grew up in rural Oklahoma, and during high school worked as a summer intern on a vegetable farm there. After college, she was looking for an internship and emailed Joel Salatin of Polyface Farm, the now-famous farmer and local food advocate was featured in the book *The Omnivore's Dilemma*. She was one of the first two female interns to work on Salatin's farm. She lived and worked for six months at Polyface, learning every aspect of poultry production. "I started out my first four days processing. I started on the gutting table. I then learned the killing," says Erica. That experience learning each station of poultry processing was pivotal to her being able to eventually start her own farm.

"Fluency is key," says Erica. "I learned all the ins and outs of poultry and everything there is to do with chickens." Having one area of expertise gave her the confidence to start a farm.

She and Joel started out with chickens

and a few cows. A friend gave them a long-term, no-interest loan to buy their first herd of cows. The idea of finding monetary help from friends and people in your community to get started is one that Erica feels is part of the communal spirit of farming. "I feel strongly that this can be an incredible way to infuse your business with capital—finding people you know to give you a low-interest loan. You write up a contract so it's legit. We've done it three times now. Sometimes all you need is a thousand bucks. We couldn't have bought a herd of cows off the bat without that help."

But even with this financial help, Erica feels like the one big mistake they made was to have not started the business with more capital. "We nickel-and-dimed it for too long. We should have gotten a loan," says Erica. There's something to be said about starting small and working into a business, but it's also an easy way to overwork yourself and to burn out. There seems to be a fine line between not biting off more than you can chew with the size of your operation, while also giving yourself enough resources to create a successful business.

Free Union Grass Farm now raises sev-eral thousand chickens each year for both meat and eggs. The Hellers keep about a dozen beef cows, a handful of pigs, and a few

Erica Hellen (standing) and Zan McKenna

hundred ducks. While the ducks are not the biggest part of their business, it's what has opened doors for them at the local farmers' market and at restaurants. There was already an established poultry farmer at the local farmers' market. But when Erica approached the market managers with the idea that she could bring ducks to market, something no one else was doing, she was able to secure a spot at the market and her business has flourished ever since.

Last year, the farm business grew to the point that Erica and Joel began looking for help. "We only have one room, but really want two people, so a couple seems natural. It's always better to have two people thinking about a [farm task]." They paid for a listing on the website Goodfoodjobs.com, where they found a lot of very qualified and educated applicants. That's where they connected with Zan and Dave. After a few conversations and a Skype session, they all knew they were a perfect fit.

Zan also began her interest in farming in high school. She was inspired by reading books including *Animal, Vegetable, Miracle*, by Barbara Kingsolver, and *The Omnivore's Dilemma*. She told her dad when she was only 16 years old that she wanted to grow up to be a farmer. She took a year before college to do volunteer agricultural work in Paraguay, with AFS, the intercultural learning and international exchange program for high school students. In Paraguay, she found a mentor in a woman who worked in the local Ecology Action research garden. The woman turned over much of the responsibility of the garden to Zan and it was there that she learned a valuable tenet of any kind of farming: "That most of the efforts should go into the soil, and that calories are only the by-product."

In a now not-so-surprising coincidence, when Zan came home to the United States, she also went to college at Warren Wilson College, though not at the same time as Erica. At Warren Wilson, Zan was on the construction crew, where she learned valuable skills that a lot of young females don't get, but that are certainly helpful on a farm. She had a year in a graduate program at University of Massachusetts before deciding to pursue farming more seriously. It was at another internship where she met Dave and learned that they both shared a love of working with animals. Soon after, they found their way to Free Union Grass Farm.

What's it like to be two women on an animal farm? And how do they deal with the idea of raising animals and then having to kill them? Erica says, "It's hard. It's the right human response that it's unsettling. I first killed chickens in a mobile processing unit in Montana. It took us four to five hours to do 50 chickens." They can now process 60 chickens in an hour. "I remember that day. Like idiots we brought chicken salad sandwiches for lunch." After attempting one bite, with the smell of dead chicken on her hands, she drank a Coke instead.

Zan killed her first chicken at Free Union Grass Farm. "On the first processing day, I went home and cried over lunch. It's intense. You can't think about it too hard every time," she says. "But with chickens, they are not the most loveable creatures. And they're dying much more humanely in our hands than at the [talons] of a hawk." Hawks kill a couple of chickens a day at Free Union, and raccoons, too. Neither one is a pretty or peaceful death. "The natural way for chickens to go is not more humane than what we're doing. All things considered, it's just not that bad. And I've done it enough now that I definitely feel more comfortable."

Both Erica and Zan see a real difference in farming animals as opposed to farming vegetables, especially when it comes to the lifestyle. Erica says, "Livestock just made more sense than veggies. With veggies there's more waste." What veggie farmers don't sell within a few days usually gets composted. "We can put our leftovers in the freezer. There's less risk." It's the risk that fuels a stereotype (mostly true) that veggie farmers are more manic. And animal farmers are, as Zan puts it, "More chill."

It's the chill lifestyle that has drawn Zan and Dave to animal farming. She says, "You don't have to be the quintessential farmer rising and sleeping with the sun. We like to have a life." Working with Erica and Joel has shown them that "farming doesn't have to be as desperate as we thought. We can still go out to concerts and enjoy the things we liked to do before we farmed."

Basic Considerations for Common Farm Animals

Dogs

A dog or two should be the first beast you add to your farm. The country is full of all kinds of mostly nocturnal and opportunistic creatures, from raccoons to bears to skunks, possums to foxes and even rats. These creatures search out humans because they know there will be food left around in some form or another. A dog, or even the smell of a dog, will keep most of these creatures from creating a nuisance around your farm.

And dogs are the best all-around guard animal for any other animal you have on the farm and for yourself. I have to be honest, I'm very nervous when Michael is out of town and someone drops by the farm unannounced. Unannounced visitors to a farm are uncommon, so when it happens, it's a bit unsettling. Having my 60-pound dog, Casey, around sure gives me more peace of mind.

But dogs can also create havoc on a farm. Do not allow your friends to bring their dogs to your farm if you have animals they can harass. You never know how a dog will react to chickens roaming your yard or a horse at your fence. If a friend does bring a dog, you should not allow it out of the car without a leash, at first. You should then personally lead it around your yard and introduce it to your animals. You will very quickly learn whether it is an aggressive dog that could cause harm to your animals. If it is, put it in a pen or back in the car. You cannot train a dog in a short period of time not to chase chickens.

You also never know how any dog you buy or adopt will react initially to other animals. This is why I suggest getting a dog before getting other animals and making sure you train it properly. But that's not always possible. Once you have other

animals, I would suggest only taking a new dog as a puppy. That is the only sure way for your dog to adjust properly to the other animals—before it can do harm. If you have no alternative and end up with a dog that kills chickens, you should invest in proper fencing for the chickens and dog unless you are confident in your own ability to train.

One good method of introducing a dog to your farm and other animals is to tie the dog to you while you go about your chores. You, of course, need to be careful that your dog is not truly dangerous, first. But if you bring your dog with you while you feed the chickens, sheep, and goats and jerk its leash when it fixates or lunges at other animals, you can train it effectively. It may take time, perhaps even weeks. But the dog should eventually take your lead and remain calm around the other animals, as long as you remain calm.

When a lot of people think about dogs in the country, they assume they are just living loose and free around the property and taking care of themselves mostly. This is a recipe for trouble. Dogs can get themselves into all kinds of trouble while roaming about. When we first moved out to the country, we had no fence. Our dogs would take off and roam the county for hours at a time. They brought back deer carcasses, no doubt harassed other people's animals, and one of them was eventually hit by a car (he lived). Even in the country, you should have a way to keep your animals close to home.

At our farm, we have fenced in 2 acres around our house. The dogs, cats (most of them), and chickens live in this area with us. Outside of the fence are the donkeys and llamas in their own enclosed pasture. You will be surprised what a better feeling it is to know your dogs are enclosed around your house, where they are most needed and safe. A good option in the country is the invisible fence. These can be put up over large areas. The dogs have a collar that gives them a mild shock if they reach the perimeter. Before you judge this as cruel, you only need to think about the alternative, which is your dog being filled with porcupine needles, shot by another farmer, or hit by a car on the rural highway.

True guard dogs are not pets. If you are going to use a dog as a working guard dog, you need to find a breed specific to this purpose—and then treat that dog as another of your livestock. It lives outside with the other animals. A good book to use in choosing and training a true livestock guardian dog is *Livestock Guardians*, by Janet Vorwald Dohner.

Cats

First and foremost, cats are predators. This is both their best and worst quality as a farm animal. We have six cats at the moment and I can tell you we definitely do not have a rodent problem. However, I shudder to think of all the dead birds, bunnies, squirrels, frogs, lizards, and snakes I've found lying around the yard or stuck in the mouths of our cats.

There's probably a good balance between the number of cats you keep to handle a rodent problem and there not being too much collateral damage. But we couldn't totally control how many cats we have, as several have just shown up and decided

to stay. This is very common in the country. If you spot a cat hanging around, the first thing you should do is trap it. Feral cats that are not neutered or spayed will breed and multiply very quickly. We have friends who work at a nonprofit spay and neuter service who regularly are called to farmhouses that have been overrun by 20 cats or more because the owner let them breed uninhibited. So, use a Havahart trap, catch the cat in question, and take it to a vet to be evaluated. Oftentimes there are cheap or free spay and neuter services in a rural area that can make it less of an economic burden.

As I mentioned before in this book, cats that aren't born and raised in the country are susceptible to meeting a premature end. If you move to a farm and have a city cat, keep that cat as an indoor cat. A mature cat that had a previous life in the city may not survive as an outdoor cat on a farm for more than a few months.

Chickens

Chickens are the easiest animal to keep. If you have a yard or pasture for them to graze, you don't even have to feed them much, except maybe in the winter. They do need water and protection from predators. Every situation is different and you'll never know whether you have a predator problem until one arises. Except for our initial loss of Ted the rooster and maybe one other chicken that was stolen out of a hole in the chicken wire of our coop, we've not had much of an issue with predators. But our friend just a few miles up the road has had two full flocks completely massacred at different times by raccoons, and another time he lost several to a barn owl that had figured out how to get inside the coop from the roof. So, make sure the coop where they sleep at night is sealed up like Fort Knox and check it often for small holes in the wire.

Chickens can be kept in a coop at night and let out to roam during the day. They can also be rotated on pasture, as I

described earlier in this chapter. If you have them around your house, they will roam your yard or their chicken yard during the day and they will go into their coop at night. You can then go every night and lock them in and every morning and let them out. Initially, you may have to wait for them to bed down at night somewhere else and you can easily pick them up and put them in their coop. It will only take a night or two before they learn to do it on their own. It's not practical to lock pastured chickens in their coop every night. For them, you should use electric netted fencing to keep them safe and/or leave your guard dog out in the pasture with them.

Do you need a rooster? Not necessarily. Chickens will lay eggs without a rooster; they just won't be fertilized and cannot hatch out as chicks. But we've always had a rooster and we've never lost a chicken in broad daylight, even to a hawk. Roosters typically have very bad dispositions. You would, too, if you were in charge of protecting a bunch of defenseless chickens and you were blind at night. Roosters have a job to do and they do it well.

I do not believe in trimming chicken beaks or cutting their wing feathers. If you do not overcrowd your chickens, you should not have any problems with their eating eggs. And if you do, I think it's more humane to find the culprit and cull it than cut off the beaks of all your chickens. And why keep them from flying if they want to? If you have trees in your yard, it is a natural thing for a chicken to roost and will protect them from some types of predators.

Unless you are keeping chickens for meat, do not allow their eggs to hatch. The typical ratio of roosters to hens in hatched eggs is about 1:1. The first and only time we allowed our chickens' eggs to hatch, we ended up with seven roosters and two hens. The only thing to do then is send off the roosters to end up in someone's pot. Of course, if you are keeping chickens for eggs and meat, then you can have a never-ending source of both by continually breeding and processing them, all the while eating lots of fresh eggs. They are an efficient and cost-effective source of protein.

Just don't try to make any money selling eggs. It's a sideline business and it takes thousands of chickens to make the profit anywhere near worthwhile. However, getting into the chicken meat business is now quite lucrative. Learn the trade first as an intern on another farm. It takes a lot of killing, plucking, butchering, and freezing to make your living this way. But people are now willing to pay top dollar for all-natural, humanely raised chicken and it can certainly be a viable business as part of a diversified farm.

Llamas and Alpacas

"Do you use their fleece?" That's usually the first question I'm asked when I tell people I have llamas. Actually, the fleece of llamas is not high quality and is difficult to spin into yarn. Llamas have guard hair and it's very tedious to separate it from the usable wool before you spin it. In fact, llamas were bred as pack animals and even as meat. Alpacas, on the other hand, are bred for their wool. It's finer quality and can bring top dollar. I wish I'd known that before buying llamas. But I probably would have done it anyway because Rico and Ferdinand, our two llamas, are cute and amusing and make us smile.

We keep llamas as pets and for their manure—remember llamanure? It's a terrific, well-balanced fertilizer that you can put directly on your plants. Llamas are also guard animals. Like other guard animals, they are most effective when you do not socialize with them. If you have alpacas, which are much smaller and susceptible to being attacked by predators, you should have a llama or two that lives with them, to serve as guardians.

Llamas and alpacas are very easy to take care of. They do not need costly feed and depending on where you live, you may have to feed them little. In Virginia, our llamas graze on their own from spring to late autumn. In the winter, for two llamas, we use about 15 square bales of hay and three bags of llama food. That's less than $100 in food per year.

The care of both llamas and alpacas consists of shearing them in the spring so they don't overheat in the winter, trimming their toenails (each foot has two toes with nails, and the bottom of their foot is a pad like a dog's) every six weeks or so, and deworming them with a shot every six weeks. We also have the vet out once a year to give them their annual vaccinations, which costs about $75. They are hardy animals and heat is more dangerous to them than snow. They lived wild in the mountains, after all.

Horses and Donkeys

If you're a horse person, you're probably horrified that donkeys are lumped in here. And as a donkey person, I almost feel the same. Horses and donkeys, after all, are completely different species. Horses are the divas of the animal world—expensive, temperamental, and prone to injury. Donkeys, or long-ears, are brilliant, hardy, and pretty much take care of themselves. Or maybe that's just the donkey lover in me coming out. That said, there's certainly a connection between women and horses. And if you feel that connection, then horses may be in your future.

In reality, the care of each is not that much different, except for the cost. The real determination on cost for a horse comes with the pedigree. If you're buying an expensive racing, jumping, or riding horse, you are going to spend not only on the initial price of the animal, but on the very high-quality feed it needs, the top-notch veterinarian you need, the expensive stable you need to keep it in, the quality farrier that will keep its hooves trimmed and healthy, and all the various saddles, bridles, blankets, and on and on. Of course, there are variations on this, from the old beat-up farm horse that doesn't need much in the way of care or regular attention, to the primadonna horse described above.

Donkeys, on the other hand, have been said to be able to live on air, meaning they require very little in the way of quality pasture or food. They are desert animals and can survive with little food and water. In fact, they are very susceptible to becoming obese in many parts of this country. We feed our two donkeys about the same as we feed our llamas—about $100 worth of hay and other food per year.

Unless you are hooking a cart up to them or some other indignity, there really is no equipment you need to own donkeys. For ours, Tyler and Whipper, we have two halters with leads, some tools to trim their hooves and that's pretty much it. They only see a vet about every three years. We paid $600 for both of them 11 years ago.

Why have donkeys? Donkeys are like having big dogs. They are definitely good guard animals and keep stray dogs away and their personality is infectious. They also provide us with manure for our flowers.

And if you have horses, it's a good idea to have a donkey. Horses and donkeys aren't always best friends. But donkeys have been used for hundreds of years to be companion animals to horses, to guard them and keep them calm. Donkeys are the unsung heroes of many a horse farm operation. Just don't tell the horses that.

Cattle

Cattle are one of the few farm animals that don't make good pets. Most are very large and not all that friendly. They are, of course, wonderful animals that you will appreciate after some time. But the sheer amount and weight of the feed and water it takes to keep them healthy and growing should be a consideration for any farmer, woman or not.

You will also need special holding pens and facilities for them. They need to be corralled, restrained, and vaccinated often. If this thought is worrisome, then cattle may not be for you. On the other hand, cattle can be set loose for much of the time if you have healthy pasture. And you only have to worry about their water source, in this case. You can buy two-year-old steers (castrated males) in the spring and raise them for about two years before sending them off to be processed.

Processing is where the real expense comes in. There are fewer and fewer processing facilities in this country because of consolidation in the beef industry. So, you may have to drive your cattle hundreds of miles to be processed. Then you have to go back to pick up the meat. Or you hire someone else to do this. It all amounts to your profit's going directly to the oil industry.

This is not to say that you shouldn't consider raising cattle. Perhaps you've inherited hundreds of acres of land from a family member. Cattle can be the ideal way to create a productive system for paying the property taxes yearly on that property. If you have land of this size, it makes economic sense for you to have a farm manager who would do all the legwork necessary to take care of a large number of animals. Women all over the Midwest are inheriting vast tracts of land and putting them to similar use.

What about having a cow for milk? A cow will continue to give off milk for about a year, give or take a few months, after giving birth. So, you will need to continue to breed your milk cows to keep the milk flowing. But keep this in mind: Milking twice a day, every day × 365 days = 730 milkings/year.

Because of all of these considerations, a beef farmer and a milk farmer are two of the best friends you can have. Why not grow vegetables and flowers and buy or trade your meat and milk with the farmer down the road?

Goats and Sheep

Goats are wily as a coyote and sheep are a lot more intelligent than some would leave you to believe. Both can live on less-than-quality pastures and do well in rocky areas. They both are gentle creatures, small enough to pick up yourself in most cases, and can steal your heart.

The primary concern with both animals is keeping them contained and protected from predators. Train them quickly to come to you by bribing them with food. Use moveable electric fencing. When housing them in stationary, nonelectric fencing, make sure it is locked tight. Put all supports on the outside of the fence, as goats can use them to climb and sheep can injure themselves on them. Make sure the fence is at least 5 feet tall. Now would be a good time to get a llama for protection, too.

Parasites are a big problem with both sheep and goats. And the parasites can become resistant to treatment. But there's an easy way to manage this, by using the FAMACHA chart, to which you can compare the color of your animal's eyes to find out its anemia levels. It allows you to identify only the animals that are in need of treatment. This saves money and helps prevent parasite resistance. It's recommended that you use it only if you are properly trained by a veterinarian. But we know farmers who have obtained the chart online and use it without incident. One farmer recommends you check your animals once a month; treat those chemically that rate 3, 4, or 5 on the chart; and give the rest a draft of garlic and cider vinegar, which is a natural way to keep parasites in check.

Whatever the reason you are keeping either animal, whether it be for meat, wool, or milk, you should be trained by another farmer. Milking, shearing, and butchering are specialty arts that require training and practice.

Pigs

The biggest problem with pigs is that you will fall in love with them. And if you have children, they will fall in love, too. But a pig can grow to 500 pounds in a year and will be up at 700 pounds or more before you know it. When this happens, the meat is no longer useful and you're stuck feeding the equivalent of an army unit or wasting the animal's meat. So, before you get a pig, steel yourself to the idea that it will be a summer romance only.

With this in mind, the only practical way to keep and raise pigs for the beginner is to buy sucklings in the early spring and butcher them in the fall. Of course, you can keep a sow and breed yourself, but that's a whole 'nother can of worms. Sows

are extremely protective of their young. And when the time comes to separate the sow from the piglet, there will be hell to pay. Many farmers have been maimed or killed in this way. Until you know more about it, stick with the raising of seasonal pigs that you keep for only a season.

Pigs don't require a lot of space. They like to be kept in wooded areas, where they can root around the roots of trees for grubs, and also in pastures. They can be pastured and put into a rotational grazing pattern, using portable electric fencing. They serve as living tillers. They don't get a whole lot of their nutrients from the actual grass, but they root up all the soil for bugs, so you need to be careful that they don't do lasting damage to your pasture. Move them often. You can also put them in a pen that does not require much space, a minimum of 100 square feet per pig. Do give them a puddle somewhere to wallow in and a dry house (similar to a doghouse) to get out of the elements.

They require somewhere in the neighborhood of 500 pounds of feed for a season. Use only organic, all-natural feeds. There is no need to dock their ears and tails. This is needless cruelty. They should be wormed about every eight weeks. But a pig is not a starter farm animal. They are destructive when not managed properly, require hundreds of pounds of feed, and they are quite dangerous. But did you see those beautiful eyelashes?

Bees

Bees are insects, not animals. But they can be livestock. Raising bees helps increase the pollination rate of your plants, while also creating a tasty by-product for your table. Bees are a valuable part of your farm ecosystem. But put aside any notion that you will make much money raising them.

The costs to set up the hives are in the neighborhood of $500 per hive. Then there's the cost of all the honey extracting equipment—another investment of about $600. And the various other safety equipment and tools is another $500 or so. So, just getting started with all you need is approaching $2,000. So far, we've never made more than about $600 in sales of our honey in any one year. And, a few years, we've had close to nothing. I would estimate it might take us a good eight years before we have a strong, profitable year. The people who can make money doing this are carpenters who make their own hives and who are also entomology experts who know how to raise their own queens.

This is not to discourage you from keeping bees. It's a wonderful pursuit that is hugely important to our environment and the future of our food system. Some say bees are responsible for pollinating a third of all our plant food. And while the recent problems with colony collapse disorder are lessening,

Bees sting only when threatened. This happens most often when you are disturbing their hive. When bees are away from their hive, they are not aggressive. The only way to get stung is if you place your hand on them directly or step on them.

there's still good reason for all of us who can to raise bees. And who knows, maybe you are that rare breed of person who will dedicate your life to propagating healthy queens and sharing them with the world.

What should you do if you want to become a beekeeper? Look up your local beekeepers' association and go to their meeting. Or find another beekeeper who is willing to mentor you. This is not an undertaking that you can learn on your own from books or the Internet. If you are allergic to bees, do not attempt to raise them. I'm stung quite often.

The general yearly makeup of a small bee operation:

- Order a queen and a nuc (small, four-frame hive of bees) from a reputable and preferably local beekeeper in the winter.

- Order your hive, which will consist of a bottom board, two deep boxes with ten frames in them each, a medium-size honey box also with ten frames, a feeder box, a top board, and a lid. You can find a full hive kit, complete with the bee suit and tools, online.

- Set up your hive (initially just one deep box on the bottom board) on concrete blocks or something else sturdy to keep it off the wet ground. Locate it in a sunny spot away from anything that might disturb it.

- When the nuc and queen arrive in the spring (via mail, usually), you follow the directions and place them in the hive. The queen stays in a cage and the bees eventually free her after they've become used to her and accepted her as their queen.

- Put the feeder on the one deep box after the queen has been freed and start feeding them with sugar water: 5 pounds of sugar to 1 gallon of water.

- After the queen starts laying and the bee population is thriving (about two weeks), add another deep box on top of the other.

- If your hive is healthy the first year, you will add the medium-size honey box, which is basically where they will put the extra honey they will try to store for winter. This is what you take off and harvest if it gets filled up and the honeycombs are capped.

- If you have a healthy hive, you harvest the honey in July. This allows the bees time to replenish their honey stores.

- When winter comes, you remove all honey boxes and leave the bees in their two deep boxes. You install an entrance reducer on the box.

- The female bees kick out the male bees, which go off to die. In the winter, the female bees form a ball around the queen in the hive to keep her warm.

- In the spring, you open the hive to make sure they survived and then you start feeding them again.

- And another season begins.

THE FARM
AS A BUSINESS

CHAPTER NINE

To paraphrase Mr. Shakespeare, "Some are born farmers, some achieve a farming career, and some have farming thrust upon them."

As you stroll the aisles of your local farmers' market, you see other women behind tables overflowing with bountiful color. Their skin is browned from the sun and their hands are strong. Their warm smile gives off a feeling of peace with their work. Even if their eyes betray a hint of exhaustion, you get the sense that it's a satisfied exhaustion. You watch them trade their physical efforts and expertise as agricultural artisans to happy customers for handfuls of cash. You can't help but feel the desire to join them in becoming fully a part of your local community and economy.

One of the most common reasons women start to grow or raise their own food is the desire to create a sustainable food system within their community. Your own farm is a place and a space that actively expresses the values that inform your food and lifestyle choices and this extends to a farm business. It is a place where you can experiment and put into practice your ideas of how things should be done, and it's a place to share and develop these ideas with others. Now that you have a grasp of how to operate a farm, grow food, and raise animals, you may decide that creating a farm business is the next logical step. Why shouldn't you become a part of your local economy?

You'll want to develop an identity for your farm, choose a specialty to focus your energies, and set up a farm business just as you have the rest of your farm—organically. By *organically*, I mean that like the plants and animals you raise, you want to start small and allow your business to grow on its own. Just as

WOMAN-POWERED FARM

pumping up your plants with artificial fertilizer is shortsighted, injecting too much money into your business to ramp up production before you have a market to sell into and before you've created a consistent product to sell will only hurt your long-term prospects. Wasting food because you've overproduced for a market that you haven't yet developed is just as bad as over-promising to suppliers and then not delivering.

My first year of flower farming was very rudimentary. I leapt into the venture with little to no experience, but with lots of energy and a great love of flowers. Inspired by Lynn Byczynski's *The Flower Farmer*, and having attended a one-day flower growing conference at Virginia State University, I bought seeds, created a planting schedule, and convinced my husband that, yes, he really did want to help me expand our garden fencing to include four long rows for flowers.

My first two years, I sold flowers at a very small county farmers' market on Saturdays, and at the two weekday markets in Charlottesville. I felt that I was working pretty hard hauling flowers back and forth and not really selling as many as I'd like (*all* of them!). A good day was when I had $100 in sales. I was still working part-time and realized I needed to fully commit to my fledgling business. The universe seemed to concur, as I was offered a permanent spot at the Saturday Charlottesville City Market for the next season. I accepted the spot, resigned from my job, and have had five wonderful seasons selling flowers and building relationships at the City Market. I eventually stopped going to the weekday markets and took on deliveries to local restaurants and special events instead. Every year, I've sold more flowers than the last.

The business has grown mostly through my exposure at the Saturday Charlottesville City Market. By being a consistent presence at the market, I've developed relationships with customers and other vendors. My regular customers are not only the backbone of my market income, but are also so loyal and appreciative that they take my breath away. A farmers' market is a community, and I feel fortunate to have developed wonderful friends whose support and camaraderie make even the wettest, windiest Saturdays worthwhile. My farmers' market booth is my best advertisement. I get about 60 percent of my special-event business (mostly weddings) from people's stopping by to admire the flowers. My website is also a great marketing tool and has brought in business, especially from people searching for local and sustainable flowers. Of course, having the most beautiful, longest-lasting flowers around is what really wins people over.

Flower farming is my third career. I'm lucky that at this point in my life I have a partner with a full-time job of his own whose income pays for the majority of our household expenses, so my business is a second income. Also, my enterprise is a small one and I have no intention of growing my business larger than I can handle alone, or with a bit of part-time help. Ted's Last Stand Farm and Gardens is financially self-sustaining and provides me with a modest income after expenses have been paid. I didn't need to start with a surge of capital to get my business off the ground. This isn't the case for many beginning farm businesses. (See page 289 for ideas about getting funding to start your farm business later.)

The only truly dependable production technologies are those that are sustainable over the long term. By that very definition, they must avoid erosion, pollution, environmental degradation, and resource waste. Any rational food-production system will emphasize the well-being of the soil-air-water biosphere, the creatures which inhabit it, and the human beings who depend upon it.

—*ELIOT COLEMAN*

The New Organic Grower: A Master's Manual of Tools and Techniques for the Home and Market Gardener

Starting Small and Discovering Your Business Specialty

Think of your business as another key link in your farm's health, just like the water, soil, and animals. The business will bring in money that will contribute to the health of the others, which in turn will bring in more money, and on and on. When starting a farm business, you should pay careful attention to your land and its resources. If your business overtaxes your land and resources, it is not sustainable. Creating a farm system in which all of the elements are healthy and work together are keys to having a successful farm business.

Working with what you have to find the specialty that will thrive is very important. Don't plan to keep sheep for their wool if you have low-lying, damp land. If your heart is set on keeping bees and selling their honey, make sure you have flowering perennials and a good water source nearby. If you have poor soil and want to raise vegetables, then you had better be ready to take the time necessary to build soil fertility (which can take years) or the resources to truck in new soil and compost to get yourself a head start on a vegetable operation. Don't try to fit your fantasy of a farm business into a piece of land that isn't right for it.

Whatever type of farm business you decide to pursue, starting small is the key and will allow you to find what you truly enjoy doing and will also allow your land time to tell you if your idea of a farm business can thrive upon it. Starting small and growing into a market is the only way to safely deliver a quality product without overinvesting money and effort. It's not surprising that the USDA's statistics show that the majority of women in farming are starting small by raising food for their family, then increasing their production to raise food for their communities through sales at a farmers' market or through a CSA.[10]

There are certainly successful farm women who have dreamed big and jumped in on a large scale. However, these women most certainly had years of farm or garden expertise and training. They probably had college-level instruction in business management. Just like a musician, no one starts out as a concert pianist. Farming is a craft and an art as well as a difficult business. Give yourself time to learn it without overexposing yourself financially or physically.

Don't take on too much all at once. Try not to take on any debt, at least at first. And if you do, always think about how much you have to grow and sell to pay for that debt. A good tomato sells for about $2 at most markets. It takes 5,000 tomatoes planted, nurtured, grown, and sold to pay for a new, $10,000 tractor. When you keep that in mind, the tractor may not make sense for you, at least in the beginning.

Another benefit of starting small is that it allows you to make mistakes. Small mistakes are wonderful opportunities to learn, and they remind us to pay attention. We have permanent raised beds for our flowers. We've fought weeds growing in the pathways for years. We've tried it all, from growing clover and rye (too much mowing and doesn't stay out of the beds), laying down black plastic (eventually breaks down and you have bits of it in your soil), laying stone over landscape fabric (the rocks make their way into your beds), and landscape fabric covered in woodchips. We've just spent a full day pulling up the landscape fabric we laid a few years back, which became overrun with weeds. Now we're stuck with a few hundred feet of nonrecyclable, shredded landscape fabric. This could have been a big problem if we'd started out with acres of production and applied this horrible idea. But because we started small with our farm business, we were able to test-drive solutions instead of create big problems.

One strategy to find what really works for you is to start out with a diverse mix of products to sell from your farm. For instance, before pursuing it as a business, you may have already been putting into practice for yourself many of the ideas we've already talked about in this book—raising vegetables and herbs, growing fruit, collecting eggs from your chickens, harvesting honey from your bees. Try to offer a diversity of products at a market or wherever you can sell them, so that you can test which is the most popular and profitable for you. And you will also find that having diversity helps create loyal customers. They may come initially for the eggs and then discover that they have to have your honey as well.

Another strategy that has produced success for women farmers I know, is to take at least a year and intern or apprentice at a farm to learn one particular aspect of it really well. Erica Hellen from Free Union Grass Farm (see page 254) interned for a year and became confidently proficient in raising chickens and selling them for their meat and eggs. She had searched out that one key skill to get started in farming and then she diversified into other areas, such as beef, ducks, and pork. Chickens are her best seller, but it's the ducks that give her the diversity that chefs and the farmers' market are looking for.

Starting small and diversifying helps you find your passion—you discover what is successful, for the space you have, for you financially, and for your own sense of value and creativity. You allow yourself to make changes to your farm operation and vision without feeling that you are wasting your previous efforts, time, and money. I think this "learning curve" is an essential part of any new venture.

Starting a Farm Business

As with any project or venture, your farm business requires you to make a plan. A business plan is a good idea, but not one I pursued initially. After having done the research into what varieties of flowers will thrive in my area, I then needed to learn which would sell in that area. That is, apart from actually growing the flowers, my business only developed after I went to market and learned what customers desired.

I started out just selling at farmers' markets. Then I branched out to florists and then eventually to doing full-service weddings. Each market has its own needs, so I had to adapt as my business grew. Florists need larger quantities of particular flowers and of similar height and uniformity. At the farmers' market, I need a larger diversity of flowers that appeal to my particular customers, and I need to have different flowers throughout the season, to keep my customers' interest. For weddings and events, I need to grow flowers of the shapes and colors that are sought after by brides, as well as plan to have enough filler and greenery for bouquets and arrangements. As I grow with the customer and market in mind, I know where, when, and to whom I will be able to sell my product. For my business, it is better for me to have more flowers than I need, rather than risk not having any when I get those last-minute event requests. My flowers are my best advertisement and I try not to miss an opportunity to get them out in the public eye.

Planning and Record Keeping

Planning for the coming season, and thinking ahead to what you'd like your business to look like in two, three, and five years, takes careful thought, self-evaluation, research, and good record

FARM BUSINESS KEYS TO SUCCESS

- **Quality:** Another reason to start small. Focus on doing a small amount at the very highest level. You have the upper hand on large growers, in this respect.

- **Consistency:** Plan your crops and products so you can offer them for the maximum time during the season.

- **Quantity:** Have enough product to show some bounty. An overflowing table of tomatoes sells a lot better than one basketful. Eventually, quantity means you can expand your customer base to other businesses. But don't sacrifice quality.

- **Diversity:** Before you specialize, give people as many opportunities as possible to buy from you. You may eventually become known for one specialty, but until then, building a customer base is the key.

- **Nonperishables:** It's very difficult not to waste some amount of any product you sell if it's perishable. Diversify your offerings so that if you have troubles with weather or disease, you can fall back on something that doesn't require being sold immediately. For instance, along with my flowers, I sell beaded jewelry I create over the winter, books, T-shirts, dried herbs, and various other crafts. They are sidelines that add a good 25 percent to my market sales and offer diversity to my customers. And they don't spoil from week to week.

keeping. Many websites offer information and help. Beginning Farmers (www.beginningfarmers.org) is one of these sites; it offers a variety of resources about planning and training, as well as the legal aspects of setting up your farm business. Your local agricultural extension agency is also a good place to find information about how to start a farm business in your particular state and county.

There are necessary steps to establishing your farm as a business.

Be sure to check with your county office to make sure you are following all the proper procedures.

- Register your farm as a business with the county.

- Establish your farm's legal structure (sole proprietor, partnership, corporation, limited liability corporation, cooperative)—sole proprietor is usually enough at first, until you grow to a level where you may be taking on more risk and debt. Consult with an attorney for the best option.

- Register with the US Department of Taxation to obtain a Federal Employer Identification Number, and register with your state's Department of Taxation as well, so as to file and pay all the necessary taxes for your business.

- Obtain the necessary licenses, permits, and certifications for your type of business and the products you intend to sell.

- Open a business bank account.

- Seek out a reliable accountant and lawyer who are familiar with the particulars of farm businesses.

- Obtain the necessary insurance for property and liability.

- Record keeping is the next consideration. Devise a method for keeping track of all of your expenses and sales (save your receipts!). This is not only necessary for tax purposes, but is also the key element in growing and streamlining your business. Take a look at the Farm Income and Expenses tax form, schedule F. This will help you categorize your income and expenses so that filing your federal income tax won't seem so overwhelming—you can file your receipts into these categories throughout the year.

Keep track of what you sell, to whom you sell it, and what prices you are charging for the product. I use an Excel spreadsheet and separate my sales monthly by Farmers' Market, Deliveries, and Special Events. I also keep a written record of what flowers I am selling, in what quantity, and at what time of year. This makes it clear for me which flower crops are actually making money, as well as which crops I've grown successfully throughout the different months of the season. Use spreadsheets!

This can seem like a lot of information to try to make sense of, but you don't have to reinvent the wheel. There are many helpful websites, complete with charts and guiding questions, to help you put together a record-keeping system that works for you. An excellent resource is *The Organic Farmer's Business Handbook*, by Richard Wiswall. Wiswall provides a comprehensive guide for financial and crop record keeping that shows how to make your farm profitable. Depending on the size of your farm, you may want to purchase business software, such as QuickBooks, or take advantage of some of the online business applications, such as Square.

Debt and Financing

While I shy away from taking on any debt and would advise you to do the same, at least in the early stages of your farm business, depending on the size and type of your farming operation, you many need to find funding to help you produce the best product you can. In fact, Erica Hellen of Free Union Grass Farm (see page 254) believes she nickel-and-dimed it for too long at the beginning. A nice bit of capital would have eased those initial years when burning out on farming was a real possibility. Erica has since taken low-interest loans in small amounts, not from banks, but from people the Hellers know in the community, to increase the number of cattle they raise for their grass-fed beef.

Sometimes, going into debt is necessary. It takes time to build a farm, establish a business, and build a customer base. Every business grows to a point where there are two options: add more financing and grow to the next level, or scale back and keep your income and work level at a lower level. At Forrest Green Farm, Krista Rahm hosts whole living workshops; runs a farm store stocked with her own herbal teas, remedies, and spices; and has homeschooled her two children. She and her husband, Rob, also raise and sell free-range chicken and their eggs, as well as growing and selling over 200 varieties of plants. Forrest Green Farm looks and feels like a little piece of paradise, and it is a very successful farm business. To make this happen, they had no choice but to take on some debt, but they have a proper perspective on their efforts. They know they aren't going to get rich farming, but they are doing what they love in a place they want to be, and their farm makes you want to be a part of it, too. That is, they took on debt to continue their desired way of life, not just to inject capital into their business. And because the farm is profitable, they are finally beginning to catch up and pay it off. In this way, debt certainly seems worth it, as long as it's manageable.

Gail Hobbs-Page of Caromont Farm (see page 292) ran a Kickstarter campaign to raise the money for her state-of-the-art milking system and upgrades to her sterile cheese-making facility. Crowd-funding of this type has become a boon to local farm and food producers. But while it seems like an easy source of free money, a lot of work and preparation must go into it. You need to develop a full campaign; you must provide giveaways and incentives for contributors, and you need to manage all the social media available to get the word out, as well as perhaps producing a video to explain your goals (see Sources of Funding, page 289). For a short time, it's like running a completely different business. So, you should go into it knowing it will take all of your energies for a while.

As Ron Macher states in *Making Your Small Farm Profitable*, "The reason to make your farm profitable is simple: As long as you take in more money than you spend, your farm can improve in fertility and has a potential for sustainability. A farm can generate large numbers of bills—your goal is to be able to pay them."

Sources of Funding

- **Friends and family:** Just be sure that you are clear if/when they will be paid back. Will you pay them in cash or in products from the farm? Creating a written contract is suggested.

- **Farm Credit Network:** A nationwide network (www.farmcreditnetwork.com) of borrower-owned lending institutions and specialized service organizations. Search out your local office and find a woman loan officer.

- **Local banks and credit unions:** They are more receptive than the big banks to small farm businesses. But search out, by word of mouth, a loan officer (again, why not a woman?) who has experience in farm businesses.

- **United States Department of Agriculture (USDA):** Provides loans to farmers. However, they have very specific eligibility requirements and most often serve as a last resort for farmers who can prove that they are unable to get a loan elsewhere.

- **Crowd-funding:** Visit such websites as Kickstarter.com and Indigogo.com. You are required to submit an idea and then take the project to completion. It's also strongly recommended that you produce a professional video, which, of course, is an expense. And if your campaign does not earn the amount of your goal, all of your efforts will be for naught. So, this should be for a business that already has an identity and a following on social media.

- **Community Supported Agriculture (CSA):** You've probably heard about CSAs. A farm offers "shares," to individuals or families, of its seasonal produce. The farm gets a payment up front at the beginning of the season and then the shareholder gets an allotment of food from the farm each week throughout the season. This serves a short-term loan that allows the farm to scale at the beginning of the season before its products are ready for market. This, too, is for the experienced farmer, as you are making promises to people and taking their money against what you can deliver to them in the future. The worst thing that can happen to a farmer's reputation is to shut down a CSA after having spent all the members' money.

Customer Service

To run a farm business successfully, you need to take care of your customers, not only the farm.

CONSISTENCY AND RELIABILITY

This means being at the same place at the same time, every time. People need to know where and when to find you and they need to know that you will be there. If you run a farm stand, keep consistent hours. If you sell at a farmers' market, try to get a permanent location so that you can be in the same spot every week, and show up every week. This element of customer service also applies to returning e-mails and phone calls. If you make people wait, chances are that they will move on to the next option.

PRODUCT QUALITY

Give your customers a clear and quantifiable reason to buy your product rather than someone else's. If I sell my flowers based on their freshness and long vase life, this had better hold true every time the customer purchases flowers from me. And if not, I will replace them for free. While I sincerely believe my flowers are the most beautiful at the market, this is subjective and comes into play with marketing strategies. Another objective way I can set my product apart in terms of quality is that I don't use any pesticides, herbicides, or synthetic fertilizers—my farm is pollinator friendly. Find specific, objective qualities about your product that differentiate you from the competition, and make the customers aware of these differences.

MEETING THE CUSTOMERS' NEEDS

The most successful CSA farm program in my area set itself apart from the competition by allowing its subscribers to choose their vegetables, fruits, and flowers rather than being handed a presorted allotment. Listen to your customers and pay attention to the details that make a difference to them. Because my booth at the farmers' market is at one of the corners, it is a logical first and last stop for many customers. They will come to my booth first and choose their flowers, and then I'll hold their purchase in a bucket of water while they do the remainder of their market shopping. As they leave, I hand them their flowers, wrapped and ready to take home. In addition to bunches and premade bouquets, I also offer my customers the novelty of picking and choosing their own stems from buckets of flowers. The other flower vendors at the market don't offer this option because of the amount of damage and loss of product (flowers broken and bruised due to unskilled handling). Also, being able to accept credit cards at the market meets the need of those customers who prefer not to carry around cash, and helps pull in those impulse buys after the allotted cash has been spent.

EXPERTISE/KNOWLEDGE

Be ready to answer questions about your product, and be willing to share your experiences, success, and failures. Do your research and provide your customers with that extra bit of information. Customers at a farmers' market want to be educated. They are interested in what you do. Give them tips on how to store what they buy from you. Provide recipes and edible product. Answer their questions about their own growing efforts. Sharing your love and knowledge of your product is a sure way to make a genuine connection with your customers.

RELATIONSHIPS

Learn your regular customers' names (I write the names of good customers next to their purchases, which helps me remember

GAIL HOBBS-PAGE

Caromont Farm
Esmont, Virginia

"Women have what I call the farmer's eye. I know like a mom does, the different yells [of my goats]. I know the I'm-stuck-in-a-feeder yell, the I'm-being-attacked-by-a-coyote yell, and the I'm-horny yell," says Gail Hobbs-Page, on her goat and cheese farm in the rolling hills of Esmont, Virginia. She has a close connection to her animals and always feels that pressing need to nurture them and feed them like her own children. And women have been the key to nurturing her business and creating the successful cheesemaking operation now called Caromont Farm.

Gail's path to farming came through the food business. She started in restaurants and worked her way through some of the finest in the South, including the Magnolia Grill in Chapel Hill, North Carolina. In Tidewater, Virginia, she found her husband, who was also in the restaurant business and they set out to buy a piece of land that they could farm. She eventually ended up in the Charlot-tesville, Virginia, area in the kitchen at Hamilton's. There she became one of the first chefs to be known for working with local farmers and sourcing much of her ingredients locally.

Having grown up on a farm where she cared for cows, pigs, and goats, it was only natural for Gail to connect her roots to her current love of good food. So, she bought a couple of goats and started making cheese, mostly for herself and friends. "I was naive, but I put all my previous knowledge to work."

At about that time, the singer Dave Matthews and his wife, Ashley Harper, were making a name for themselves locally with their farm, Best of What's Around. Again, citing her naïveté, Gail approached them and asked whether they would let her work for free. They agreed and bought Gail 13 goats to manage. Along with the goats, she started a 100-person CSA vegetable subscription program. But the farm eventually began to go in a different direction and they shut down most of the business. It was through the generosity of Harper that Gail gained the spark that she needed to jump into farming for herself: She gave her the goats and Gail was on her way to becoming a cheese maker. That first small herd is now about 125 strong, of which 100 are milked every day.

Gail built her dairy on instinct, mostly. But there were mentors, too. She relied on other farmers for advice and then she crossed paths with two women inspectors from the Department of Agriculture who took it upon themselves to help her along. "They became like my soul sisters and helped me get inspected," says Gail. And when her first milk came back with unhealthy levels of whatever milk is mea-sured by, they helped and advised her how to bring it into compliance.

"At that time, most resources were for the large milk producers and not for the cheese makers. The old-time farmers didn't even know there was such a thing as goat cheese. I got really frustrated with them," says Gail. And the local suppliers turned out to be even less helpful. The feed salesmen would come to court her and laugh at what they perceived as her misguided approach to farming.

She insisted on feed that didn't have harmful chemicals and additives. Once, she conducted a web search for the ingredients on one of the labels of feed she was sold and found only websites written in Chinese. That was an "Aha" moment for her and she began demanding clean and natural food for her goats. Soon, the feed reps stopped laughing at her and it became, "You bitch." There's a fine line to dealing with the old-boy network, though. "You have to learn to stand up, ask questions, and be strong willed," says Gail. "But if you're too much of a bitch, you'll be ignored."

That first year, she produced 300 pounds of goat cheese, all from a milking cart that held only two goats at a time. She sold her

cheese, as she does now, at the local farmers' market and to local stores. She designed a logo with a friend, on the back of a cocktail napkin, and her business was born. Last year, Caromont produced over forty thousand pounds of both goat and cow cheese.

And as so many farmers learn, the first two years are usually very positive with a lot of growth. "Then in the third year, everything breaks," reports Gail. And by the fifth year, the workload had become overwhelming. That's when she hired a cheese maker and began looking for more financial resources.

The banks told her they didn't do any agriculture loans. The USDA gives loans, but only after you can show that no one else will give you one. The irony of the government's only lending to those farmers with the worst credit and highest risk of default is not lost on Gail. Finally, she found a sympathetic ear in a woman (see a trend?) loan officer at the Farm Credit of Virginia, and was able to secure $35,000.

At about the same time, her interns (she now has one full-time young woman intern living on her farm each year) had turned her on to Kickstarter, the crowd-funding website.

They talked her into starting a campaign to raise money for a new pasteurizer. She didn't know what she was getting into, as she's not well versed in social media. It took a lot of work and planning, and she hired a professional videographer who was paid a portion of the funding proceeds. In the end, she was able to raise $40,000 more from Kickstarter. She now had $75,000 to work with and only half of it was debt. She parlayed that into a new pasteurizer and other modern equipment that has now made her business profitable, and she's able to pay herself a salary.

But it's the animals that Gail appreciates most about her business. It's evident in the way they flock and play around her on the farm. Raising animals can be emotionally challenging, at least in the beginning. "You have to let things go. You have to cull," advises Gail. "You have to send your babies off to slaughter. Things die and sometimes you have to shoot things." While it's hard to prepare for this aspect of raising animals, she advises that you should be aware that there's no getting around it.

Another "Aha" moment in this regard came to Gail during a 100°F summer day when she was all alone on the farm. One of her bucks had developed fly-strike, a serious condition that results in maggots eating away at wounds on the animal. She had not acted on it early enough, the condition had developed into a brain infection and the buck was too far gone. She had to screw up her courage, shoot the buck, drag him out of his enclosure, and bury him. "I was sick about it. That animal could have blessed someone's table," says Gail. But she's now comfortable sharing this story in hopes that another woman farmer won't make the same mistake and will learn when "to let things go."

When asked about advice she would give to other women farmers just starting out, Gail offers, "Get a sense of what the work is like first. Don't romanticize it. Ask a lot of questions and be strong-willed. Women have to work harder and be smarter." She also recommends continuing education so that you can constantly improve the way you do things and learn from others. She insists, "We take ourselves seriously. We're trying to make this a business."

them), take an interest in them, and keep them informed. Greeting customers by name and remembering their preferences will not only encourage repeat business, but will also create a personal connection. Loyalty is developed not only through a special and reliable product, but also through the relationship we have with the person who provides that product. Keep your customers up to date with what is happening on your farm and with your products via e-mail, website posts, or social media.

Marketing

Marketing is all about making sure your product gets noticed by potential customers and eventually purchased.

BRANDING

The first step in this process is branding your farm business so as to tell a story, reflect your farming philosophy, and distinguish you easily from competitors. Your "brand" is the idea of your business that you want to convey to the customer. Focused around your business name and logo, your brand not only represents you and your product, from your own point of view, but how you want customers to perceive your business and your product.

The name you choose for your business is a very personal choice. With a farm business, your business name is often the name of your farm. This is the case with my farm and business "Ted's Last Stand." Yes, there is a story behind this name and when customers ask about it, I'm always happy to tell it.

Once you've chosen a name for your business, you need to connect that name to your product in a way that customers will remember. To create more recognition for your business, you need to create a logo that ties in with your business name.

Spend the effort to create something you like—you'll be looking at it a lot! Your business name, your logo, and a concise description of your product and business are all you need to get started marketing your business. It's wise to enlist the help of an experienced graphic designer.

BUSINESS CARDS

Create and have business cards printed. You can have this done at a locally owned print shop or chain, or by using one of the many online design and print websites.

TED'S LAST STAND: THE NAMING OF OUR FARM

When we arrived at our newly purchased farm, after an exhausting cross-country trip from California, we were dismayed by the amount of work the place needed. But this was offset by our delight in discovering that the previous owners had left their chickens—one white rooster and one white hen. I immediately named them Ted and Bev (they were just such a cute little couple). Ted was a handsome white Leghorn rooster, with a bright red comb. Beverly was his svelte female companion, who rarely left his side. Ted and Bev had both been spared the injustice of wing-clipping and beak trimming, so their features were fully formed and striking. Ted's spurs were formidable—at least 4 inches long and curved to sharp points, though they served only as a warning and were never used in anger.

Ted was a true Southern gentleman. His barrel chest bulged with pride as he strutted around his farm. He paid particular attention to Bev, always scanning for trouble and calling to her if she were out of his keen sight. He would crouch protectively over Bev when she was laying her daily egg, so he'd be the first to confront anyone who dared interrupt. I loved offering Ted the treat of half a grape, which he would pluck with his beak, delicately from my palm, turning to offer it to the much more human-shy Bev, then turning back to receive the second half for himself.

Ted's crow was less impressive than his demeanor—not so much a cock-a-doodle-doo as more a gurgled Rebel yell. It could be described as a cross between a war whoop and a rabbit scream made by a sick goose. But what it lacked in sonorous melody was made up with the enthusiasm of delivery, which could come at any hour from before dawn to dusk.

While we loved the idea of Ted and Bev and the novelty of a daily egg hunt, neither of us knew much about chickens. Ted and Bev had no coop to sleep in at night, as the previous owners just relied on their dogs to keep predators at bay. They hadn't even constructed a laying box for Bev, so she used a flower box outside our living room window as a substitute. Bev was able to fly into the sheltering comfort of a large holly tree at night, but Ted, too top heavy to fly, settled in at night on the front porch. We figured that they had been getting along just fine that way before, so we weren't in any hurry to improve their housing situation.

It wasn't long after we moved in that we awoke one morning, concerned because Ted had not begun crowing and the sun was now fully above the tree line. We quickly went out for a look, only to find a trail of white feathers leading from the porch, across the field and into the woods. Ted had no doubt met his demise in the mouth of a fox. And thus we learned our first lesson of farming—chickens and other defenseless farm animals will die if not protected at night.

Our guilt at not properly protecting Ted weighed on us. Bev had been laying eggs in the window box. When she became broody after the loss of Ted, I didn't have the heart to pull her off her eggs. I really didn't think they would hatch, having been left largely to their own devices for the past several days. But about three weeks later, Bev hatched out seven fluffy chicks in the window box. By that time, we had built a coop to keep Bev safe. We soon discovered that five of the seven chicks were male and only two were female.

This led us to our second lesson of farming—unless you're prepared to live with the cacophony and fighting instinct of multiple roosters, find homes or cooking pots for them all, or kill the males yourself, then don't ever allow your chickens' eggs to hatch. And so, we found another farmer who enjoyed making homemade chicken soup and we've felt guilty about that ever since.

We did keep one of the little roosters, and named him Ted Jr., or TJ as we call him. While TJ is not quite as people-friendly as his dad, he is just as much of a gentleman. He looks after the girls—we kept the two female chicks—and makes sure they've all headed into the coop at night before he heads in himself.

A couple of months after Ted's demise, when we were tossing around ideas for naming our farm, I brought up the idea of memorializing Ted by naming our farm in his honor. This is how Ted's Last Stand Farm and Gardens came to be.

Since the naming of the farm, Ted's Last Stand has also become the name of our specialty cut flower business.

By generating curiosity in the name, I've created an opportunity to connect with potential customers, and through the story of Ted, the likelihood that they will remember my business and me is high. By branding my business using Ted, the rooster, I'm conveying the idea that the qualities that Ted possessed—care, attention, courtesy—are qualities you will find in my business, as well as the fact that we care about all the living things on our farm. Not to mention that the image of a rooster is a very popular one, judging from our T-shirt sales.

SIGNS/BANNERS/STICKERS

If you will be setting up a booth in a market environment (including fairs, festivals, and expositions) you will want to have a sign or banner made to identify your business and help you stand out among the crowd. Depending on your situation, you may also consider sandwich board signs and magnetic car signs. I recommend investing in high-quality signs made by a dependable local business. My banner is now six years old and still as good as new—I was referred to the company by another farmer, and have since referred several others to the company. Stickers are also a good and cheap way to brand everything from jars of honey, to egg cartons to paper and plastic bags.

WEBSITES

Create a website. This can be a stumbling block for some of us farmers. When are we supposed to have the time to do this?! Well, you may not have the time, but you need to do it anyway. Luckily, designing and keeping up with a website has gotten easier in the past few years. Many user-friendly and inexpensive sites provide templates and hosting to create your own website.

Alternatively, you can hire someone to do it for you; I recommend getting a referral to a local web designer so that you can be closely involved with your website creation and upkeep. Your website is a place to show off your product, explain what you do, and share the work you love. It should reflect the ideas, personality, and goals of your business.

You will find that you do not have the time to update a website regularly with blog posts, and so on. A website should start out as a glorified, electronic business card. Use lots of photos and describe the various products you offer and where to find you. But at least in the beginning, make it a static site that doesn't need regular updates.

Later on, you may develop a store component to your site and sell directly from it. This can be a valuable source of additional revenue. Selling nonperishable related products—garden shears for instance—is an easy way to get a little cash flow in the off-season. You can also delve into the area of sponsored ads. But you may need to hire a website manager to manage the daily needs.

Look into how to add your farm (with a link to your own website) to online directories or advertising spaces on other websites relating to your business, such as farmers' market sites. You may be able to do this entirely for free, or as a free exchange (linking to those sites from your website), and unless your contact information changes, this requires no upkeep once it is done.

There are myriad ways to generate income from websites, but your time as a small farmer is best focused on your farm. Hire someone and offer them a percentage of online sales if you want to build this aspect of your business. My site serves mainly to attract wedding customers looking for local flowers. They find me through the site and then we connect via e-mail and phone. I only update it once a year.

SOCIAL MEDIA

Using social media (e.g., Facebook, Twitter, Instagram) is another means of sharing your business with the public and attracting potential customers. I keep my website static without regular updates. Social media is a way to keep in touch with your customers without the time and labor of website work. In this age of instant media, it is worth exploring the possible ways that different social media accounts could help promote your business and link you to others in your profession. These accounts take consistent and fairly constant monitoring and contributions of information to reap results, but can be well worth the time and effort. But when you're first starting out, focus most of your energies on your product. Social media is further down the priority list for most farmers.

Setting Prices

Pricing your product is about more than being competitive with other growers or farmers. One of the big mistakes almost every farmer makes when starting out is to underprice the products. It's very tempting to do this as you're trying to lure in customers and increase the their number, whether you are selling to farmers' markets, restaurants, or wholesalers. There's a temptation to underprice the other farmers in your market, to try to get a greater volume of customers. But you are small and you should not compete on volume. While you may get a few bargain hunters, what you're really doing is lowering the perceived value of your product. If you are offering a quality product that you're confident can compete against the competition, you should keep your prices high. There is and should be a premium on local, artisan farm products. Customers are buying more than a tomato. They are buying peace of mind that it was grown to the highest standards by a farmer that they can talk to and feel comfortable with. I've raised my prices every year and it's only resulted in making more money and selling more flowers.

Publicity

Quite different from marketing, which is more about the practicality of connecting product and consumer, publicity is not something many farmers ever dream of attracting. But publicity is good for business and it provides a level of validation to your business in the eyes of potential customers. Local newspapers and radio/television news programs are always hungry to find the next local interest story. There can't be too many of them to

fill up all the space and time these news people need to fill. So, find an angle on your farm that is interesting. The simple fact that you are a woman farmer is of growing interest to the media and the public, or I wouldn't be writing this book. Perhaps you're a one-woman farmer raising a rare breed of miniature cattle to fill the growing market for grass-fed beef. Find an angle or just tell your story and let the reporter find it for you. It's easy to locate e-mail addresses of your local news people online. E-mail them with an introduction and idea and you may be surprised at the response. Of course, you want to time this outreach to coincide with your selling season, so you can take advantage of the free advertising.

The Farmers' Market

The farmers' market is most often the primary source of sales for any beginning farmer. And some farmers are so successful at it that they make their entire living selling at several markets a week. Where your market is located and the quality of the management and customers will determine how big a part of your business the farmers' market will be.

Our Saturday market is one of the biggest in Virginia; it has over 100 vendors and thousands of people shop there each week. Over the last five years, I've developed a loyal customer base of people that return year after year. I'd estimate that more than a third of my customers each week are "regulars." And all of these customers, at one time or another, buy flowers from me outside the market.

The farmers' market is not only a good revenue source, as you are selling your product direct to the consumer, but it's also an advertisement for your farm and your products. Much of the wedding business I get comes from young couples and mothers browsing the farmers' market to see which local flower and food vendors they might use for their wedding. Many are visiting the market this season to see what might be available a year from now. Even if I don't get a sale from them right away, I may land a big wedding next year. I also get a lot of delivery and other special-event business from the market.

When you're first starting out, you may have a hard time getting a space at the market. Most of the good markets are now very competitive and space is limited. You'll have to prove yourself and your product. The first things you will want to do are (1) get in touch with the market manager and (2) satisfy all of the paperwork and other requirements. Also, send or drop off a sampling of your best product, even if the manager doesn't ask for it.

Make nice with the market manager and take some time to describe what you're doing on your farm and what you think you can bring to the market. Go to the market and observe. Find out how many other vendors are offering products you would like to sell. If there's a lot of competition already, find a niche that isn't being filled and try to fill it. Market managers are under a lot of stress from farmers all the time. There's not a farmer in the world that has never had a complaint for a market manager. The manager is tasked with filling every spot with a wide diversity of products to keep the customers coming back, while also keeping the needy farmers happy. So, if you can show that you are easy to get along with and you are offering a product that fills a need, then you have a leg up. As we learned from Erica Hellen at Free Union Grass Farm, while her primary product was chickens, she only secured a permanent space at the market when she started offering ducks, which no one else was selling.

You might have to start out for a while as an "alternate" vendor. The key is to try to get in every week and be persistent (with a smile). When you aren't able to get a spot, go to the market anyway as a customer and make sure to say hello to the

WOMAN-POWERED FARM

market manager and express your interest in being a vendor the next week. Don't get discouraged.

Once at the market, focus on having an attractive setup with clear signage and pricing. People don't like to guess at what they are about to buy. Remember, some people may not be familiar with all the vegetables or other products you might offer; signage identifies what is for sale and makes them feel more comfortable buying and asking questions. Use table-cloths. They look more inviting than bare market tables. And try to create levels in your display. The more you can bring your products closer to eye level, the more customers will stop and look.

Bring only your best products. You gain nothing by selling anything that might turn out to be disappointing to a customer. And show that you have bounty. Instead of spreading out your products on a table, fill up baskets or pile them up. Bountiful displays are inviting. As you sell your product down throughout the day, combine your containers to keep your display looking full. For instance, if you have two baskets of potatoes, and you sell both of them down halfway, it's best to combine them into one big, bountiful basket of potatoes.

Keep smiling. It will be difficult some days when sales are slow and the sun is hot. Customers will ask ill-informed questions, complain about your prices, and tell you the other farmers have better produce and visit all kinds of other indignities upon you. But part of retail success is engaging every customer as if he or she might eventually be your best. Our friend Susan Parks at Broadhead Mountain Farm has a trick for herself that she learned from her mother: Whenever a customer says something offensive or political that you do not agree with, just smile and say, "So true." It helps to repeated it twice just for emphasis, "So true. So true."

Best Practices for the Farmers' Market

- Befriend the market manager. Gifts of produce or other products are not out of line.

- Show up at every market, rain or shine. Do not take a week off unless it's an emergency or rare vacation.

- Create a farm sign that you hang on the front of your tent or behind your table, with your farm name, logo, and where your farm is located.

- Create a display that is a representation of your farm's all-natural approach and that includes photos of your farm. Tablecloths, apple crates, and sandwich boards are a nice touch.

- Provide easy-to-read labels that describe your product with prices that can be read from outside your booth.

- Always have enough change. Only rookies run out of change.

- Accept credit cards. It's now easy to do with a smart phone, and the fee for each purchase is reasonable. We gain hundreds of dollars in business a week by accepting credit cards.

- Bring only your best products to market.

- Engage and educate your customers about your product and your approach to farming. Provide recipes on cards if you are selling food.

- Create a bountiful display, with multiple levels, and combine your products as they sell down.

- Learn your "regulars'" names.

- Make friends with other farmers—we have a lot to learn from one another.

- Keep your display clean and uncluttered. Tablecloths help hide your empty boxes under the tables.

- Attach your farm brand to every purchase that leaves your booth—stickers are a good way to seal bags or label jars if you don't want to spend money on branded packaging. Customers will be buying from many different vendors. The chance of their remembering where they bought that delicious kale is improved if you've labeled their purchase.

- Accommodate special requests without overcommitting. It's OK to say no sometimes.

- Price your product high enough that it carries perceived value. Don't get into price wars with other farmers. "Upsell" by offering a slight discount for buying more (e.g., one eggplant for $2 or three for $5).

- Find the empty niches in the market and fill them.

- Bring a diversity of products, including nonperishable items that you don't have to pick, clean, and harvest each week.

- Bring plenty of business cards and try to get the contact information of anyone who takes one.

- Have fun!

Weddings are now a big part of my overall farm business.

Growing Beyond the Farmers' Market

Eventually your business and your volume may grow to a level that the farmers' market should not be your only or primary source of sales. Many farmers get to this point and they begin selling to grocery stores or other retailers, restaurants, and wholesale businesses. They may hire someone to sell for them at the farmers' market, to free up more time for them to manage the farm and find more customers. Even if the farmers' market is still your primary source of sales, you should diversify your customer base just as you've diversified your farm. A bad week at the market can be offset by a good week of selling to local restaurants.

Restaurants

The good thing about restaurants is that they will pay a price that is higher than what you'd get from a wholesaler. It won't be as high as you can get selling direct to a consumer. But restaurants need food and some need flowers weekly and all year round. And with the current trends all pointing toward more local and sustainable foods, chefs are having a hard time finding enough local produce to satisfy the demand.

Chefs and restaurant managers are a temperamental bunch, not to be too stereotypical about it. They're almost as bad as farmers! But seriously, their expectations are very high when it comes to food. And they are used to uniformity in the product they buy from big wholesalers. They also need produce at all times of year, well out of what may be your growing season. It can be difficult to meet all their needs and sometimes their demands. You should not overpromise or get carried away trying to change your business model to meet one customer. But if a dozen chefs around town are asking for local basil all year round, then it may be time to invest in that greenhouse and provide what the market is asking for.

And like your customers at the farmers' market, even chefs need to be educated. They are certainly proficient in the preparation of food, but you'd be surprised at how many are not informed about all that goes into growing all the foods they use. It's important to fill the chefs in on all that you do to make your product excellent. As you probably notice on more and more menus, the farms and the methods of growing particular ingredients are printed right there next to the dishes. So, make sure you educate the chefs as to why your product will fit right in and help drive their sales.

Retail Businesses

Similar to dealing with restaurants, selling to retail can be both rewarding and challenging. You will not receive as high a price for your product as you would from a restaurant because retail businesses need to mark the price up, sometimes by 100 percent, to make it economically feasible. But the volume of business can be much more than a restaurant.

One of the challenges to selling in this market is that most often there are very specific packing guidelines, as required by the market or even the USDA. Depending on what you're selling and to whom, you might have to provide small portions, enclosed in clamshell packaging, with specific labels to that store and perhaps even a barcode. The extra time and packaging cost, coupled with the lower price you can get from the market, is often not worth it for the small farmer. The volume can be great, but this is an area that you should save for when you've outgrown other markets. That said, markets do buy some things in bulk that don't require packaging, so it's worth educating yourself to the local markets' needs and pursuing it if you have bulk product you'd like to sell.

Wholesale

Selling wholesale is a bit counterintuitive to the small, sustainable farmer. The idea of focusing on quantity over top quality is not why you're interested in farming, I'd imagine. But there are now local regional wholesalers of food dedicated to sourcing local products and selling them to the local market. We have one right here in our area, called the Local Food Hub; it is a nonprofit organization dedicated to providing local and sustainable food to our area markets, restaurants, and schools. Because it is nonprofit, it can offer better pricing to farmers. It also can handle many of the packaging needs of some of its customers.

These regional wholesalers are an important key to maintaining local and sustainable agriculture. So, while the big wholesale food companies are probably your last resort as a small farmer, regional wholesalers dedicated to local foods should be supported when it makes sense for your business. Even if you aren't planning to sell to them anytime soon, it's a good idea to talk to them and find out how to become a farm vendor to a wholesaler. You never know when you'll be awash in blueberries and don't have enough of your own customers to buy them.

Best Practices for Selling to Restaurants, Retail Businesses, and Wholesalers

- Bring only your best products.

- Uniformity is important. Try to bring them your product in a consistent size and shape.

- Find out when their slow time is and solicit business only then. Restaurants typically are slow before noon. Retailers and wholesalers generally are slow in early afternoon.

- Ask for a price higher than you can imagine they will accept. They will always want to bargain with you, at least initially. So, go in high and bargain down. It's almost impossible to raise prices later unless the market is so hot that they have no choice but to pay what you're asking.

- Always bring extras with you when making deliveries. And bring other varieties and products. Often you'll find that your customers underestimate what they need or they decide they'd like to try something new.

- Don't overpromise. Nothing will lose a customer faster than not delivering an order on time and to specifications.

- Search out farmers that sell to the business already and are not offering a competitive product. They are usually happy to offer advice and might even make the initial contact for you.

Interns and Apprentices

As your business grows, it is inevitable that it will be too much for one person, or even two people, to handle. Bringing in some extra hands is usually necessary. Interns or apprentices are one of the most common ways to address the need for extra help. In Chapter 2, I discuss the value of interning to learn about farming. Most of the medium-size to larger farms in my area have at least two or more interns who live and work on the farm property throughout the growing season. Just as there are drawbacks and cautions associated with being an intern, the same is true of hiring an intern.

Some of the same challenges apply to hiring regular employees. However, there are many more legal and tax considerations for regular employees than there are with interns. These vary from state to state and depend on whether the employee is full-time or part-time. I'd suggest that you consult an attorney and your accountant before hiring employees.

One of the common practices in all of farming is to use unpaid interns as free labor. This is not only illegal, it's ethically dubious. The chief concern of internships is education. Yes, you will benefit from an intern's labor, as farming is labor and any education about it requires labor. And the skills and knowledge you give to an intern certainly carry a monetary value.

The US.Department of Labor provides the following five criteria that must be applied when making the determination of whether an internship can be unpaid:

- The internship, even though it includes actual operation of the facilities of the employer, is similar to training that would be given in an educational environment.

- The internship experience is for the benefit of the intern.

- The intern does not displace regular employees, but works under close supervision of existing staff.

- The employer who provides the training derives no immediate advantage from the activities of the intern; and on occasion its operations may actually be impeded.

- The intern is not necessarily entitled to a job at the conclusion of the internship; and the employer and the intern understand that the intern is not entitled to wages for the time spent in the internship.[11]

There are obvious gray areas here. I would never personally use an unpaid intern. I feel that there will always be work on a farm, even in the context of an internship, which should be compensated with money. You will have to make your own determination. But make sure to follow the guidelines above and you can steer clear of any labor troubles.

The same places I suggested to find internships in Chapter 2 apply to where you should go looking for interns for your farm. Some key considerations and things need to be understood between you and your intern or apprentice. The clearer you are from the get-go, the better the relationship you will have and the more productive your intern and the internship will be.

WORK EXPECTATIONS

Before hiring any intern, you should clearly spell out in writing the exact nature of the jobs the intern will be expected to do and the hours he or she is expected to work. This should be written in a way that illustrates the work as educational, which it of course is.

EDUCATION

Because education is a legal requirement and in the spirit of internships, you should also spell out in writing what the intern will learn. Offer a curriculum of sorts that you will work through with the intern on a daily basis. This can, of course, include work that teaches skills of the farm internship the individual signed up for. Also, ask the intern to provide in writing what he or she hopes to learn from the experience.

HOUSING

There are no set requirements here. I know many an intern who has lived in a yurt for the summer or camped out for much of it. As with the work, you should be clear and honest about the living conditions and what you will provide in the way of food, heat/cooling, water, toilets, and showers.

COMPENSATION

There are ways you can get around paying your intern, but why? Even if you put a premium on your knowledge, you will never get a quality effort out of anyone you ask to do your work for free. This is America, after all. Pay as much as you can afford for your business and be honest with your intern about what you can afford. If you cannot afford to pay much, or even if you can, you should compensate with perks, such as free food, free lodging, or free beer. It's not a typical job, for sure. But it's also a job that's somehow more personal and connected to the spirit of your business. So, live up to that spirit and compensate your intern honestly and fairly.

INSURANCE/HEALTH

As was suggested earlier on finding a farm, you need to make sure your insurance covers employees. Typically a farm rider on your homeowner's policy will cover this. But contact your insurance agent to make sure. The agent can also provide you with a liability release form, which you should certainly have your intern sign. Make sure to have a proper first-aid kit, fire extinguishers, and other necessary emergency equipment, and train the intern in what to do in case of emergency.

COMPATIBILITY

If you can't hang out socially with your intern, find another one. You can never know for sure if you don't meet the person first, but several phone calls or Skype sessions will give you a good idea. In-person interviews are always preferable. Your intern will not learn from you and you will not gain much benefit from your intern if you are not compatible and do not see eye to eye. While it is certainly an employer/employee relationship, the most valuable internships are those that are collaborative. You may have just as much to learn from an intern as an intern has to learn from you.

HOW MANY INTERNS?

I've heard from some farmers that couples are the only way to go. They can get along in close living situations and they are in tune with each other. Even if they aren't a couple, having more than one mind focused on a project or a problem is always a good thing. They can entertain each other as well. So, if you can handle the time, pay, and housing of more than one intern at a time, then that seems preferable.

FARM EDUCATION AND FARM SCHOOLING

CHAPTER   TEN

Farms are rich ground for learning—for both adults and children. Susan Wise Bauer of Peace Hill Farm and Peace Hill Press, a publisher of homeschooling curriculum, says that she hopes the farm "instills in my kids that farming is just part of survival. Not something separate. You use food; you don't waste it." She grew up being farm schooled and has now farm schooled her four children as well. Farm schooling is only different from homeschooling in that a farm offers a wealth of opportunity to learn responsibility and to learn out in nature as well as what you'd learn in a classroom. "Regular homeschoolers have almost too much control over their schedules," says Wise Bauer. Farm schooling offers chores that teach responsibility. It doesn't matter if you want to do them or not. They've got to be done.

Teaching kids how to feed themselves and how to live in a community responsibly is the center of an education.

—ALICE WATERS

"One of the benefits of farm schooling is the life skills," says Wise Bauer. "Kids get a sense of what it's like to have chores; what it's like to take care of a living creature. It doesn't matter if the weather's terrible. It doesn't matter if you're sick. It's a sense of connection with the realities of survival."

Many women I've talked to who are returning to farming are doing it for their children. Farming offers a way for families to stay home and spend more time together. Farms offer families shared experiences. They also offer an educational alternative for children who do not fit in to the regular school experience.

Susan Wise Bauer and her brother were both farm schooled in the '70s and '80s, long before it was common. Her mother was a teacher and taught them to read before they went to regular school. Because they had such high reading skills when they went off to kindergarten, they did not fit into their class levels. They were well ahead of others in reading. Moving them ahead several grades didn't work because they had not developed in other ways. So, farm schooling was the only real option.

"We spent a lot of time doing farm chores. We had three hundred chickens. I hated those chickens so much. We ate a lot of our own animals. We had pig kills and put up pork for the entire year. At the time we didn't think of farming as educational. It was survival. I keep coming back to that word—survival," says Wise Bauer. "It was a good experience for me and I wanted to duplicate it for my own children."

Krista and Rob Rahm, of Forrest Green Farm, had a son, Dylan, with learning disabilities. After putting him in a private school and driving him over an hour round-trip each day, they realized they really couldn't justify the expense. "We didn't think we were going to be homeschoolers. In fact, we really resisted it at first," say Krista. "But the cost [of private education for Dylan] would have forced both of us to work off the farm."

So, the choice to farm school Dylan was easy. He grew up on the farm and loved to be outside. "Because he was so comfortable outside, we would go outside to read. Math happened outside with the plants," says Krista.

Their son flourished in this environment. They taught him to nature journal, using plants and bugs. Geometry was learned one season in the garden. Laying out rectangular and circular beds and calculating the growing spaces of each connected in a way that paper and a compass could not. When a huge snowstorm struck and collapsed the roof of the barn, it offered the opportunity to figure out the water weight of snow. Krista says it made Dylan's education "more real." And so when their daughter, Shannon, came of school age, there was no question she would be farm schooled.

In cities, community centers are turning to urban farming and gardening not to only offer fresh and local food for residents, but also to serve as an educational tool for all their residents, young and old. Recently, the *New York Times* talked to people at 16 different urban farms and community gardens in the city alone that they found were predominantly run and staffed by women. One farm out in East New York actually employs and teaches high school students throughout the growing season. It's not full-time farm schooling, but a supplement to their regular education.

Learning about business and how to manage money are other benefits of a farm education. Rob and Krista's daughter, Shannon, started doing markets for Forrest Green Farm when she was seven or eight. She makes felted crafts to sell and she's saved a lot of money. "Shannon's a ruthless businesswoman," Rob says with pride. And Dylan raises cattle and has bought himself a Jeep with the proceeds.

It's not just children learning at Peace Hill Farm or Forrest Green Farm. Part of both of their farms' incomes come from classes that they offer to adults. People of all ages are hungry to learn the skills of survival on a farm. At Forrest Green, they offer an intensive "Whole Life" class that focuses on raising and using herbs as well as many other basic skills of farming. They make enough money from their farm classes for adults that they no longer have to sell at the farmers' market each Saturday. At Peace Hill, Wise Bauer operates a bed-and-breakfast that offers a homesteading package. It turns out that people are eager to pay money to experience all the chores around a farm. Women are natural educators. So it makes sense that women are now finding a way to incorporate education and farming.

DEBORAH GREIG
AND HER ASSISTANT,
STELLA HAIR

East New York Farms Project
Brooklyn, New York

The clanking rumble of the elevated train running just out back of the East New York Farms project is a reminder that we're still in the city. The garden, where you can find rows of produce growing on almost a full city block in Brooklyn from March to November each year, is teeming this afternoon with high school students harvesting for the farmers' market coming this weekend.

Part of the United Community Centers, the farm was started in 1998 by a neighborhood organization and in cooperation with the Cornell Extension Agency. Growing and selling produce at its own weekend farmers' market to people in the community, the farm also employs and educates 33 students under the age of 18. All the proceeds go to maintaining the program.

Deborah Greig, the full-time farm manager, and wrangler of all the young volunteers, says, "We serve marginalized East New York.

Thirty-five percent of people living here are recent immigrants. And out of one hundred and fifty thousand residents, one in three is under the age of eighteen."

Many of the new immigrants coming to the neighborhood are from the Caribbean. As a result, the East New York Farms project grows food that serves community tastes: lots of peppers, bitter melon, long beans, and callaloo, a leafy green made into a cooked dish of the same name. Because the local markets don't offer up quality fresh produce and certainly not Caribbean specialty crops, there's always a considerable line at the weekend farmers' market booth.

Greig worked part-time farming in college, where she got the bug. She wanted to be a teacher at one point as well. But she got interested in working with young people through farming instead. "It's easy to talk to kids when farming about all kinds of subjects," she says. And it's easy to imagine that, working side by side with young people and sharing the same tasks, conversation would come more easily than in a school or typical adult/kids setting.

And it was during one of her first farm jobs that she realized her calling. "I was in the middle of hundred-foot rows of carrots on my knees. And I said to myself, 'I feel totally at peace and relaxed right now.'" She knew then that farming was her future.

She pursued formal education in agriculture from the Center for Agroecology & Sustainable Food Systems at the University of California, Santa Cruz. The program is one of the few formal training grounds in the United States that works to advance food justice. Then she returned to Brooklyn, where she grew up, and began working at the East New York Farms Project. That was eight years ago.

You can see in Greig the connection between nurturing plants and nurturing young minds. But it's not at all easy. While we were talking, more than one young person approached her about stopping work early. She called one of the leaders over and said the team needed to come up with a strategy to finish the work. The vegetables needed to be picked and sorted before they could leave.

I found the farm assistant, who was just out of college, on her knees harvesting a patch of spinach. Shella Hair, who grew up in the neighborhood, has only been farming for a year. After college, she joined AmeriCorps and helped build the first urban farm in Red Hook, Brooklyn. That's where she got the farming bug.

"It's [the idea] of taking a vacant lot or space that no one uses and building a garden or farm that makes me love urban agriculture. This lot was trashed and dirty with debris and rubbish. And now it's a farm where residents of East New York can come and get fresh produce and just have an alternative to what they can eat in the community. Not just what they can buy in the community," says Hair.

She also likes the attention it brings from her neighbors when they see her walking down the street in her farm clothes or when they walk by and catch her kneeling in a bed of dirt and vegetables. "The residents come past and they're like, 'Good job, Stella!' I love the environment and the atmosphere. It makes me feel good. I love it," says Hair.

As she harvested more spinach, she continued, "You think of the country down South where my grandparents were raised and my great grandparents . . . picking cotton. So, it's like, 'A farm in the city? A farm in Brooklyn?' It gives you that feeling of, Wow, I'm a farmer in the city. My community, they don't know too much about it. But I'm trying to bring it to them. They're not convinced yet. But they will be."

And even the kids who are working on the farm and are getting paid aren't yet convinced. When asked, none of them wanted to eventually be a farmer. They all said the work was too hard. They were just hoping to make some money. But as they stood, side by side, picking long beans from the trellis and talking and joking with each other, their smiles betrayed a peace and calm that they may not get at school. And it wasn't hard to imagine that a seed had already been planted and one day, even if these kids didn't grow up to be farmers, that they would be crucial to the future of bringing quality, local foods to places like East New York.

ADVICE FOR WOULD-BE FARM SCHOOLERS

- Parents starting out always say, "How do I know I'm not going to miss anything? Well, you are going to miss something. But it will be okay."—Susan Wise Bauer

- Get online and start researching. A great place to start is the homeschool website and forum www.welltrainedmind.com.

- Buy a full curriculum in a box. This is everything you will need: lesson plans, books, tests, etc. There are many options online, depending on the type of education you want for your child, both secular and religious. Once you start working your way through it, you will figure out the areas you are confident in teaching and those for which you will need guidance.

- Libraries are rich resources for homeschool information and librarians can point you in the right direction.

- Find a mentor who is doing the same thing or has homeschooled her children. You need someone you can call for advice. There are many homeschool networks that you can connect with online. Home-school co-ops are now common as well. Co-ops have a teacher or mother proficient in a subject, who teaches it for a fee. For instance, for chemistry, your kids may go to a co-op class to do their lab work and then write up their lab report and take their tests from home.

- Parents second-guess themselves as schoolwork at home can go very quickly. But if you're doing your tests and your kids are doing well on them, then you're doing enough.

- Do as much education outside as possible. Nature journaling and field tripping are a great way to take advantage of an unconventional education.

- Don't get too stressed about keeping to a schedule. Farm schooling is often done in the evening, after chores are done.

- Let your children find a farm business or part of the farm they can manage as their own. This includes making money from the proceeds.

You don't have to have children yourself to foster education on your farm. It's important for all of us who are gaining the benefits of farm living to educate the next generation about where their food comes from and how they might contribute. Farm tours are a good way to do this while making a little extra money for your farm. Many schools and homeschool networks are eager to find field trip opportunities. And schools especially have budgets for these. Charging a tour fee, say $5/child or a flat fee of $50 or $100 for school or family tours, is perfectly acceptable. You can of course offer tours for free, but one of the things that everyone needs to be educated about is that farming is real work and farmers need to be paid fairly for their time.

RESOURCES

Ashworth, Suzanne. *Seed to Seed*, 2nd Edition. Decorah, IA: Seed Savers Exchange, 2002.

Berry, Wendell. *Home Economics*. New York, NY: Farrar, Straus and Giroux, 1987.

Birutta, Gale. *Storey's Guide to Raising Llamas*. North Adams, MA: Storey Publishing, 1997.

Bradley, Ellis, and Martin. *The Organic Gardener's Handbook of Natural Insect and Disease Control*. Emmaus, PA: Rodale Press Inc., 2010.

Bradley, Ellis, and Phillips. *Rodale's Ultimate Encyclopedia of Organic Gardening*. Emmaus, PA: Rodale Press Inc., 2009.

Bubel, Nancy. *The New Seed-Starters Handbook*. Emmaus, PA: Rodale Press, 1988.

Byczynski, Lynn. *The Flower Farmer*. White River Junction, VT: Chelsea Green Publishing, 2008.

Childs, Laura. *The Joy of Keeping Farm Animals*. New York, NY: Skyhorse Publishing, 2010.

Coleman, Eliot. *Four-Season Harvest*. White River Junction, VT: Chelsea Green Publishing, 1999.

Coleman, Eliot. *The Winter Harvest Handbook*. White River Junction, VT: Chelsea Green Publishing, 2009.

Damrosch, Barbara. *The Garden Primer*. New York, NY: Workman Publishing Company, Inc., 2008.

Damerow, Gail. *Storey's Guide to Raising Chickens*. North Adams, MA: Storey Publishing, 1995.

Deppe, Carol. *The Resilient Gardener*. White River Junction, VT: Chelsea Green Publishing, 2010.

Dohner, Janet Vorwald. *Livestock Guardians*. North Adams, MA: Storey Publishing, 2007.

Fortier, Jean-Martin. *The Market Gardener*. Gabriola Island, BC, Canada: New Society Publishers, 2014.

Fukuoka, Masanobu. *The One-Straw Revolution*. New York, NY: New York Review Books, 1978.

Gowdy-Wygant, Cecilia. *Cultivating Victory*. Pittsburgh, PA: University of Pittsburgh Press, 2013.

Grandin, Temple. *Animals in Translation*. New York, NY: Harcourt Publishing, 2005.

Grandin, Temple. *The Way I See It*. Arlington, TX: Future Horizons, Inc., 2011.

Horwood, Catherine. *Gardening Women*. London: Virago Press, 2010.

Jeavons, John. *How to Grow More Vegetables*, 8th edtion. Berkeley, CA: Ten Speed Press, 2012.

Jensen, Joan M. *Promise to the Land*. Albuquerque, NM: University of New Mexico Press, 1991.

Kunitz, Stanley. *The Wild Braid*. New York, NY: W.W. Norton & Company, Inc., 2005.

Lauters, Amy Mattson. *More Than a Farmer's Wife*. Columbia, MI: University of Missouri Press, 2009.

Levatino, Michael and Audrey. *The Joy of Hobby Farming*. New York, NY: Skyhorse Publishing, 2011.

Logsdon, Gene. *The Contrary Farmer's Invitation to Gardening*. White River Junction, VT: Chelsea Green Publishing, 1997.

Looby, George. *Backyard Livestock*, revised edition. Woodstock, VT: Countryman Press, 1990.

Lowenfels, Jeff & Lewis, Wayne. *Teaming with Microbes: The Organic Gardener's Guide to the Soil Food Web*. Timber Press, 2010.

Macher, Ron. *Making Your Small Farm Profitable*. North Adams, MA: Storey Publishing, 1999.

Magdoff, Fred & Van Es, Harold. *Building Soils for Better Crops*. Waldorf, MD: Sustainable Agriculture Publications, 2009.

Megyesi, Jennifer. *The Joy of Keeping Chickens*. New York, NY: Skyhorse Publishing, 2009.

Petrini, Carlo. *Slow Food Nation*. New York, NY: Rizzoli International Publications, 2013.

Pitzer, Sara. *Homegrown Whole Grains*. North Adams, MA: Storey Publishing, 2009.

Salatin, Joel. *You Can Farm*. Swoope, Va: Polyface Farms, 1998.

Schwenke, Karl. *Successful Small-Scale Farming*. North Adams, MA: Storey Publishing, 1991.

Storey, John and Martha. *Storey's Basic Country Skills*. North Adams, MA: Storey Publishing, 1999.

Sussman, Julie & Glakas-Tenet, Stephanie. *Dare to Repair*. New York, NY: Harper Collins, 2002.

Thomas, Heather Smith. *Storey's Guide to Raising Beef Cattle*. North Adams, MA: Storey Publishing, 2009.

Thurkettle, Vincent. *The Wood Fire Handbook*. London: Octopus Publishing, 2012.

Ussery, Harvey. *The Small-Scale Poultry Flock*. White River Junction, VT: Chelsea Green Publish, 2011.

Virginia Whole Farm Planning: An Educational Program for Farm Startup and Development. Niewolny, Kim and others. September 16, 2014. Publication ID: AEE50P

Wilson, Gilbert L. *Native American Gardening*. Mineola, NY: Dover Publications, Inc., 2005.

Wiswall, Richard. *The Organic Farmer's Business Handbook*. White River Junction, VT: Chelsea Green, 2009.

Resources — Internet

American Agri-Women
www.americanagriwomen.org

ATTRA resources for beginning farmers
attra.ncat.org/attra-pub/local_food/startup.
html

Beginning Farmers
www.beginningfarmers.org

British Columbia Farm Women's Network
www.bcfwn.100mile.com

Caromont Farm
http://www.caromontfarm.com/

Connecticut Women's Agricultural Network
www.ctfarmrisk.uconn.edu

Earth Tools, Inc.
http://www.earthtoolsbcs.com/

East New York Farms
http://www.eastnewyorkfarms.org/

Ecology Action
http://www.growbiointensive.org/

Fairweather Farm
http://www.fairweatherfarmers.com/

Fedco Seeds
http://www.fedcoseeds.com/

Food Tank
www.foodtank.com

Forrest Green Farm
http://www.forrestgreenfarm.com/

Free Union Grass Farm
http://www.freeuniongrassfarm.com/

Green Heron Tools, LLC
www.greenherontools.com

Growing for Market
www.growingformarket.com

Herb Angel
www.etsy.com/shop/VaHerbAngel

Holistic Management International
www.holisticmanagement.org

IRS, Publication 225, Farmer's Tax Guide
www.irs.gov/uac/
Publication-225,-Farmer's-Tax-Guide

Johnny's Selected Seeds
http://www.johnnyseeds.com/

Midwest Organic and Sustainable Education
Service (MOSES)
www.mosesorganic.org/projects/
rural-womens-project/

New England Small Farm Institute
http://www.smallfarm.org/

North Carolina Women of the Land Agricultural
Network   www.ncwolan.org

Oregon State University Women's Farmer
Networks   www.smallfarms.oregonstate.edu/
women-farmer-networks/main

Peaceful Valley Farm and Garden Supply
http://www.groworganic.com/

Penn State Extention
http://extension.psu.edu/business/
ag-alternatives/farm-management/
starting-or-diversifying-an-agricultural-business

Pennsylvania Women's Agricultural Network
(PA-WAgN)
www.agsci.psu.edu.wagn

Plate to Politics
www.platetopolitics.org

Rodale Institute
http://rodaleinstitute.org/

Rural Development Leadership Network
www.ruraldevelopment.org

South Carolina Women's Agricultural Network
(PA-WAgN)
www.clemson.edu/public/ciecd/focus_areas/
leadership/programs/scwc/

United States Department of Agriculture
http://www.usda.gov/wps/portal/usda/
usdahome

University of Vermont Women's Agricultural
Network   www.uvm.edu/wagn/

Well-Trained Mind
http://www.welltrainedmind.com/

Women's Agricultural Network: University of
Vermont
www.uvm.edu/wagn/

Women, Food and Agriculture Network
www.wfan.org

ENDNOTES

1   Michael Tortorello, "In the Garden: Mother Nature's Daughters," *New York Times*, September 28, 2014, http://www.nytimes.com/2014/08/28/garden/mother-natures-daughters.html?hpw&rref=garden&action=click&pgtype=Homepage&version=HpHedThumbWell&module=well-region&region=bottom-well&WT.nav=bottom-well.

2   Gowdy-Wygnant, Cecilia. *Cultivating Victory: The Women's Land Army & the Victory Garden Movement* (Pittsburgh, PA: University of Pittsburgh Press, 2013).

3   Christopher Doering, "Breaking the 'Grass Ceiling': More Women Are Farming," *USA Today*, March 17, 2013, http://www.usatoday.com/story/news/nation/2013/03/17/women-farmers-increasing/1993009/.

4   Joan M. Jensen, *Promise to the Land* (Albuquerque: University of New Mexico Press, 1991), 2.

5   Doering.

6   Eric Anthamatten, "What Does It Mean to Throw Like a Girl?" *New York Times*, August 24, 2014, http://opinionator.blogs.nytimes.com/2014/08/24/what-does-it-mean-to-throw-like-a-girl/?module=Search&mabReward=relbias%3Aw%2C%7B%221%22%3A%22RI%3A10%22%7D.

7   Marsha Wright, Disinfecting a Domestic Well with Shock Clorination, Guide M-115 (Las Cruces, New Mexico State University, http://aces.nmsu.edu/pubs/_m/M115/welcome.html.

8   Elaine R. Ingham, Soil Food Web, http://www.nrcs.usda.gov/wps/portal/nrcs/detailfull/soils/health/biology/?cid=nrcs142p2_053868.

9   Jeff Lowenfels and Wayne Lewis, T*eaming with Microbes: The Organic Gardener's Guide to the Soil Food Web* (Portland, OR: Timber Press, 2010).

10   Shenoy, Rupa, "Q&A: How Women are Taking a Leadership Role in Agriculture," Minnesota Public Radio, April 28, 2013. http://www.mprnews.org/story/2013/04/28/women-leadership-farming.

11   US Department of Labor, Wage and Hour Division, Fact Sheet #71: "Internship Programs Under the Fair Labor Standards Act," http://www.dol.gov/whd/regs/compliance/whdfs71.htm.

INDEX

accountants, for farm
businesses, 286
acidic soils, 166–67
acreage considerations
for farms, 59, 61
for kitchen gardens, 155–56
active farm business, defined, 62
aerated compost tea, 173
AFS (formerly American Field
Service), 32
alkaline soils, 166–67
alpacas, 252, 265
amount of land. *See* acreage
considerations
anaerobic compost tea, 173
Animal, Vegetable, Miracle (Kingsolver), 256
animal feed, 252
animal manure, 233–34, 265
animals. *See* farm animals; and
specific animals
Animals in Translation (Grandin),
22, 237, 246–47
annuals, 159–60
Anthamatten, Eric, 79
Apiaceae, 189
apprenticeships, 28–35, 283,
314
Angel Shockley's experience,
30–31
how to find, 35
resources, 32
artisan, use of term, 12
ash pans, for stoves, 98
Asteraceae, 189

attire. *See* safety attire
augers, 133, 147

back posture, 78–81
backhoes, 147
bacteria tests, for wells, 91–92, 94
bacteria treatments, for septic
systems, 95
banks, as farm funding source,
289
banners, 300, 307, 308
barbed-wire fencing, 127, 240
barns, 58
basic country skills, 109–51
basic engine maintenance,
118–19, 120–21, 122
basic farm gear, 83–85
bee stings, 274
beef, 231–32, 268–69
bees (beekeeping), 273–75
acreage considerations, 61
general yearly makeup of
operation, 275
Beginning Farmers, 285
bench grinders, 144
Bermuda (wire) grass, 57
Berry, Wendell, 158
Best of What's Around Farm, 293
Beverly (chicken), 235–36, 238,
298–99
biointensive agriculture, 32, 163
bleach
for disinfecting wells, 93–94
in septic systems, 95
Blecha, Elizabeth, 13

blowers, for stoves, 98
board fencing, 127, 133
body, posture, position, and
breath, 78–81
books, recommended, 26,
331–32
boots, 84, 85, 110, 118
bottle jacks, 126, 151
branding, 296–97, 308
Brassicaceae, 188, 189, 192
breathing, 80
breeding animals, 252–53
broadcast seeders, 142, 210
broadforks, 139
Broadhead Mountain Farm
(Keswick, VA), 72, 74, 79, 307
buckwheat, 177, 178–79, 180
Buffalo-Bird Woman, 19
bug spray, 85
bungee cords, 151
bush hogs, 147
business bank accounts, 286
business cards, 297, 309
business debt, 69, 282, 287
business farms. *See* farm
businesses
business loans, 289, 294
business name, 296–97
business plans, 284
business taxes, 62–63, 286
buying a farm, 52–65
acreage considerations, 15,
59, 61
author's experience, 37,
39–40

benefits of, 43
Camey Stewart's experience, 44–45
insurance, 63, 65
location, 58, 156
potential of the land, 53–58
questions to ask, 40–42
taxes, zoning, and easements, 62–63
Byczynski, Lynn, 279

canola oil, 114
Caromont Farm (Esmont, VA), 288, 292–95
catalytic combustion, 96, 97
cats, 234–35, 261–62
cattle, 252, 268–69
acreage considerations, 61
processing plants, 231–32, 238–39, 268–69
Cedar Ring Greens (Frankfort, KY), 149
ceiling fans, 105
Center for Agroecology & Sustainable Food Systems (CASFS), 32, 325
chainsaw oil, 114
chainsaws, 110–14
general considerations, 111
safety attire, 110
starting, 114
step-by-step instructions, 112–14
Charlottesville City Market. *See* Saturday Charlottesville City Market
Chenopodiaceae, 189
chicken coops, 249, 262–63
chickens, 231, 237, 262–64
processing plants, 231–32, 238–39
rotational grazing, 248
wrangling, 250

chimney, 98–99
cleaning out, 100, 102–3
chimney sweeps, 98–99, 100
classes, 26, 87, 241
clothing, 81–85. *See also* safety attire
cold climates, winterizing, 105
cold frames, 157, 213–14
step-by-step instructions, 208–9
Coleman, Eliot, 280
colony collapse disorder, 273–74
community gardens, 65–66, 69
community support, 87, 89–90
Great Donkey Escape, 225–29
compensation, for interns, 316
compost (composting), 168–76
basic setup, 171–72
three-pile system, 170–71
compost tea, 172–73
basic recipe, 176
step-by-step directions, 174–75
conventional rows, 214
cotyledons, 200
cover crops, 57, 177, 180
credit cards, 308
credit unions, as farm funding source, 289
creosote, 99
crop names, 189
crop rotation, 163, 188, 192, 219
vegetable families, 189, 192
crowbars, 136–37
crowd-funding, 288, 289, 294–95
CSAs (Community Supported Agriculture), 50, 289, 290–91
Cucurbitaceae, 189, 192
customer service, 290–91, 296

Damrosch, Barbara, 153

Dare to Repair (Sussman and Glakas-Tenet), 90
debt, 69, 282, 287
deworming, 241, 245, 265
diamond sharpeners, 143
digging bars, 136–37
Dirty Life, The (Kimball), 77
disks, 147, 214
dogs, 236, 246, 258–60
Dohner, Janet Vorwald, 260
donkeys, 238, 267–68
fencing, 130, 133
Great Donkey Escape, 225–29
as guard animals, 234, 246
drills, 144
drip irrigation, 156, 165, 214–17
raised beds, 160, 162
step-by-step installation instructions, 216–17
drought, 56, 165

ear protection, 110
earthworms, 54, 166
easements, 63
East New York Farms Project (Brooklyn, NY), 324–26
Easy2Start, 114
Ecofan, 96, 98
Ecology Action, 32
eggs, 231, 237, 263–64
electric air compressors, 123
electric fencing, 128–29
electric generators, 101, 104–5
electric netting, 127, 129
enclosures for animals, 249. *See also* fencing
engine maintenance, 118–19, 120–21, 122
engine startup, 122
environmental easements, 63
equipment. *See* tools; tractors
Ericaceae, 189

estate planning and taxes, 46
eye protection, 110

Fabaceae, 188, 189, 192
Facebook, 301–2
Fairweather Farms (Afton, VA),
  71, 184–87
FAMACHA chart, 271
farm, finding. *See* finding a farm
farm animals, 221–73. *See also*
  specific animals
    acreage considerations, 61
    breeding, 252–53
    enclosures for, 240, 249
    grazing, 247–48
    Great Donkey Escape,
      225–29
    life cycle of, 238–39
    parasites, 243, 245
    partnership with, 231–34
    size of, 239–40
    socialization, 245–47
    veterinarian/first aid,
      240–41
    wrangling, 249–50
farm businesses, 277–317
    acreage considerations, 59, 61
    customer service, 290–91,
      296
    debt and financing, 69, 282,
      287–89
    keys to success, 285
    marketing, 296–302
    planning and record keeping,
      284–87
    pricing products, 302, 307,
      313
    publicity, 302, 304
    restaurant supply, 310–11,
      313
    retail supply, 311–12, 313
    starting small and diversifying,
      281–83

taxes, zoning, and
  easements, 62–63
farm classes and workshops,
  26, 87
Farm Credit Network, 289
farm history of women, 19–21
Farm Income and Expenses Tax
  Form (Schedule F), 286
farm loans, 289, 294
farm safety, 71–85
farm schooling, 319–29
Farm Seed Comparison Chart,
  178–79
farm structures, 58
farm tours, 23, 329
farm trucks, 150–51
farmers' markets, 277–80, 304–9
    best practices, 308–9
    buying land nearby, 52
Federal Employer Identification
  Number, 286
feed/supply stores, 53
felling wedges, 111, 113, 118
fencing, 58, 127–33, 240
feral cats, 262
filters, 90, 105, 122
financing farm businesses, 287–89
finding a farm, 37–69
    acreage considerations, 59, 61
    author's experience, 37,
      39–40
    buying, 52–65
    Camey Stewart's experience,
      44–45
    economics of, 41–42
    insurance, 63, 65
    leasing, 46–47, 49–51
    potential of the land, 53–58
    questions to ask, 40–42
    taxes, zoning, and
      easements, 62–63
    urban farms, community and
      school gardens, 65–69

firewood, 99, 107, 111
    chainsaw operation, step-by-
      step instructions, 112–14
    splitting, 115, 116–18
first aid, 240–41
first aid classes, 241
fixed panel fencing, 132
flat tires, 123, 126
    step-by-step instructions
      for fixing, 124–25
fleas, 243, 245
flood zones, 56
Flower Farmer, The (Byczynski),
  279
flowers (flower farming), 14, 56,
  181, 279–80, 284
    acreage considerations, 61
    annuals and perennials, 159–60
foot (hoof) health, 252
footwear, 84, 85, 110, 118
Forrest Green Farm (Louisa, VA),
  287–88, 321–23
Fortier, Jean-Martin, 162
4-H, 23, 32
four-stroke engines, 119
Free Union Grass Farm (Virginia),
  254–57, 283, 287, 305
freedom, and farming, 12–13
friendships, 78
front-end loaders, 146, 147
frost protection, 210–14, 219
funding farm businesses, 287–89
fungi, 166, 172–73
furnaces, 105
Future Farmers of America (FFA),
  32

garden centers, 196
garden hoses, 165
garden journals, 153, 155
garden layout, 159–62
Garden Primer, The (Damrosch),
  153

garden rows, 157, 159–60, 162
   building tips, 182, 183
   planting out seedlings, 203,
     206
   starting seeds, 206–10
garden shears, 134, 135
gardening tools, 134–44
gas storage, 119
generators, 101, 104–5
genetically modified organism
   (GMO), 195
"girlie throw," 79–80
Glakas-Tenet, Stephanie, 90
gloves, 118
goals, 41
goats, 239–40, 270–71
Good Food Jobs, 32
Good Life, The (Nearing), 25
Goodfoodjobs.com, 256
government subsidies, 42
Grandin, Temple, 22–23, 223,
   224, 237, 246–47
grazing, 233–34, 247–48
Great Donkey Escape, 225–29
Green Heron Tools, 80, 135
greenmarkets. See farmers'
   markets
Greig, Deborah, 324–26
groundwater, 162, 165. See also
   wells
guardian animals, 234, 246, 260,
   265

Hair, Stella, 324–26
hand sharpeners, 143
hand tools, 134–44
hard woods, 99
Harper, Ashley, 293
harvest knives, 134, 135
hats, 85
hay, acreage considerations, 61
health insurance, for interns, 317
healthy farmers, 71–85

heat mats, 199, 200
heirloom seeds, 195
Hellen, Erica, 71–72, 254–57,
   283, 287, 305
Herb Angel (Charlottesville, VA),
   30–31
herbicides, 218
high-tension wire fencing, 130
   step-by-step installation
     instructions, 140–41
Hobbs-Page, Gail, 253, 288,
   292–95
hoes, 137–38
homeowner's insurance, farm
   rider, 63, 65
homeschooling, 321–23, 328
Honey Creek Angus
   (Fredericksburg, TX), 44–45
hoof health, 252
hoop houses (high tunnels),
   212–13
horses, 238, 246, 267–68
   fencing, 127, 130, 133
   foot health, 252
housing, for interns, 316
hybrid seeds, 195
hydroponic greenhouses, acreage
   considerations, 61

inheritance, 21, 46
injuries, 71–72, 74
   keys to avoiding, 78–79, 81
   posture, position, and
     breath, 78–80
Instagram, 301–2
insurance, 63, 65
   for interns, 317
internal parasites, 243, 245
Internal Revenue Service, 286
Internet resources, 26, 52, 332–33
internships, 28–35, 283, 314
   Angel Shockley's experience,
     30–31

five criteria for, 315–17
   how to find, 35
   resources, 32
invisible fences, 260

Jensen, Joan M., 20
Johnny's Seeds, 194
joint stress, 72, 74
jumper cables, 151

Keller, Helen, 17
Kickstarter, 289, 294–95
Kimball, Kristin, 77
Kingsolver, Barbara, 256
kitchen gardens. See vegetable
   gardens
kneeling position, 81, 85
kneepads, 72, 74, 81, 85

Lamiaceae, 189
land
   amount of, 15, 59, 61,
     155–56
   potential of the, 53–58
landscape staples, 211
lawnmowers, 119
   basic maintenance, step-by-
     step instructions, 120–21
   gas storage, 119
lawyers, for farm businesses, 286
layout of garden, 159–62
leases (leasing), 46–47, 49–51
legal structures of farm businesses, 286
Lemley, Connie, 149
Let's Move!, 66
Lewis, Esther, 20
library resources, 26
licenses, 286
life cycle of farm animals, 238–39
Liliaceae, 189
Livestock Guardians (Dohner),
   260

living trusts, 46
llamas, 82, 234, 238, 265
    fencing, 130, 133
    foot health, 252
    manure (llamanure), 233–34, 265
    wrangling, 249–50
location of farm, 58, 156
logging chaps, 110
long-sleeved shirts, 85
long-term leases, 47, 49–51
low tunnels, 212

McGhee, Bradley, 227–28
Macher, Ron, 288
McKenna, Zan, 254–57
Making Your Small Farm Profitable (Macher), 288
manure, 233–34, 265
Market Gardener, The (Fortier), 162
market managers, 305, 307, 308
marketing, 296–302
markets. *See* farmers' markets
Matthews, Dave, 293
mauls, 115, 116
mechanical splitters, 115, 118, 147
mentorships, 238, 293, 328
metal thickness, of stoves, 96
milk cows, 221, 269
mindfulness, 71, 74–75
mineral rights, 62–63
mulches (mulching), 77, 180–81
muscle pulls, 72, 74, 80, 81
muscles, posture, and position, 78–81

naming farm businesses, 296–97, 298–99
National Organic Program (NOP), 195
National Sustainable Agriculture Information Service, 29, 31, 32

National Weather Service, 78
Native American Gardening: Buffalobird-Woman's Guide to Traditional Methods, 19
Native American women, 19
Natural Resources Conservation Service (NRCS), 45
Nearing, Helen, 25
neighbors, 87, 89–90
    Great Donkey Escape, 225–29
New Organic Grower (Coleman), 280
newspapers
    classified ads, 52
    publicity, 302, 304
nitrogen, 167–68, 169, 180
nurseries, 198
    acreage considerations, 61
nylon rope, 151

Obama, Michelle, 66
oil filter wrenches, 123
oil filters, 122
Omnivore's Dilemma, The (Pollan), 248, 254, 256
orchard acreage considerations, 61
organic certification, 63, 195
Organic Farmer's Business Handbook, The (Wiswall), 287
organic farming, 12, 32, 51, 63, 166, 195
outdoor faucets, 105

parasites, 243, 245, 271
Parks, Susan, 72, 74, 79, 307
pasture, 247
    rotational grazing, 247–48, 249
pathways, 77, 157, 159–60, 162
    building tips, 182, 183
Peace Hill Farm (Charles City, VA), 223, 319, 323

Peaceful Valley Farm Supply, 196
perennials, 159–60
pest control, 188, 218–19
pesticides, 51, 218
petroleum rights, 62–63
pH, 166–67
phosphorous, 167–68
pick mattocks, 136
pigs, 271–73
    fencing, 132
Pilates, 74
plant cover, 57, 177, 180
plant families, 189, 192
Plant Hardiness Zone Map, 192, 193, 194
plows, 14, 147
Plummer, Dave, 254, 256, 257
Poaceae, 189
Polygonaceae, 189
post-hole drivers, 133, 140
posture, 78–81
potash, 167–68
potential of the land, 53–58
potting soil, 199, 203
power takeoff, for tractors, 144, 146
power tillers, 139, 147, 148, 149, 214
precision hoes, 137
predators, 133, 235–36
pressure tanks, 91–92
pricing products, 302, 307, 309, 313
primer buttons, 122
processing plants, 231–32, 238–39, 268–69
    Grandin and, 22–23, 223
product quality, 290
property taxes, 51, 62
property zoning, 63
pruning shears, 134, 135
publicity, 302, 304
pumps, 91

quality control, 290

Rahm, Krista, 287–88, 321–23
Rahm, Rob, 321–23
rain barrels, 101, 164
rainwater, 156, 165
raised beds, 157, 160, 162
    drip irrigation, 156, 160
    step-by-step instructions,
        190–91
realtors, 52–53
record keeping, 286–87
repair services, 90
restaurants, 310–11, 313
retail businesses, 311–12, 313
roosters, 263–64. *See also* Ted
rope, 151
Rosaceae, 189
rotary tool sharpeners, 144
rotational grazing, 247–48, 249
row covers, 210–14, 219
rural farm setting, 58, 87, 89
rye, 177, 178–79, 180

safe farming, 71–85
safety attire, 83–85
    chainsaws, 110
    splitting firewood, 118
safety quick-release knot for
    tying up animals, 242
Sagebiel, Ruben, 44–45
Salatin, Joel, 248, 254
Saturday Charlottesville City
    Market, 279–80, 290–91, 309
school gardens, 65–66, 69
seeds and seed starting,
    192–210. *See also* seedlings
    basic materials, 199
    basic terms, 195
    Plant Hardiness Zone Map,
        192, 193, 194
    planting out seedlings, 203,
        206

resources, 331–33
soil blocks, 202–3, 204–5
starting indoors, steps,
    200–201
starting outdoors, 206–10
websites, 196, 333
seed catalogs, 25, 194, 196
    Farm Seed Comparison
        Chart, 178–79
seeders, 142, 207, 210
seedlings
    basic materials, 199
    benefits of starting your own,
        198, 199
    indoor steps, 200–201
    planting out, 203, 206
    watering, 164
septic systems, 94–95
service calls, 90
sharpening tools, 143
sheep, 270–71
Shockley, Angel, 30–31
shoes, 84, 85
short-term leases, 46–47
shovels, 81, 135
shredders, 147
signage, 300, 307, 308
sink disposal, 95
siting gardens, 158–59
size
    of farm animals, 239–40
    of farms, 59, 61
    of kitchen gardens, 155–56
soapstone liners, 97
social media, 301–2
socializing farm animals, 245–47
soil blocks, 202–3
    step-by-step instructions,
        204–5
soil fertility, and buying a farm, 57
soil food web, 157, 166–68, 167.
    See also compost
    tillers and, 148

soil quality, and buying a farm,
    54
soil testing, 66, 166–67
Solanaceae, 189, 192
Southern Exposure Seed Ex-
    change, 194
sows, 271–72
splitting firewood, 115
    step-by-step instructions,
        116–18
splitting wedges, 117
standing water, 54, 56
Stewart, Camey, 44–45
stickers, 300, 309
stoves. *See* wood-burning stoves
stress, 75, 77–78
stretching, 74
succession cropping, 163
sun exposure, 56–57, 156, 248
sunscreen, 85
surface rights, 62–63
Sussman, Julie, 90

tarps, 151
taxes, 62–63, 286
Teaming with Microbes
    (Lowenfels and Lewis), 168, 173
Ted (rooster), 235–36, 238,
    298–99
three-point hitch, for tractors,
    144, 146
throw like a girl, 79–80
ticks, 243, 245
tie downs, 151
tillers, 139, 147, 148, 149, 214
tire plug kits, 123, 124
tire pressure gauges, 123, 125,
    151
tires
    fixing flat, 123, 124–25, 126
    removing, 126
toilet paper, 95
tool angle, 80, 81, 143–44

tools, 107–51. *See also* specific
 tools
topography, 56–57
T-post fencing, 130, 131
  step-by-step installation
    instructions, 140–41
tractors, 144–49, 214
  buying tips, 147
  useful implements, 146,
    147–48
  walk-behind, 149
trailers, 253
trucks, 150–51
Twitter, 301–2
two-stroke engines, 119
tying up animals, safety quick-
  release knot for, 242

Unsettling of America, The
  (Berry), 158
urban farms, 65–66, 69
US Department of Taxation, 286
USDA (United States Department
  of Agriculture), 45, 281
  internship criteria, 315–17
  loans, 289, 294
  National Organic Program,
    195
  People's Garden, 66, 69
  Plant Hardiness Zone Map,
    192, 193, 194
  Water Quality Information
    Center, 56

vegetable gardens, 153–219
  acreage considerations, 61,
    155–56
  composting, 168–76
  cover crops, 177, 180
  crop rotation, 188, 189, 192
  drip irrigation, 214–17
  efficiency and uniformity,
    157, 162

frost protection and season
  extension, 210–14
keys for maximizing growth,
  156–57
layout of, 159–62
mulching, 180–81
pest control, 188, 218–19
plant families, 189, 192
planting out seedlings, 203,
  206
seeds and seed starting,
  192–210
site selection, 158–59
soil food web and, 157,
  166–68
watering, 156, 163–65
veterinary care, 240–41
Victory Gardens, 20, 41

walk-behind tillers, 148
walk-behind tractors, 149
Warren Wilson College
  (Asheville, NC), 32, 256
wartime, 9, 11, 19–20, 41
water (water supply), 54, 56,
  62–63, 65, 156. *See also* wells
  testing, 91–92, 94
water rights, 62–63
water table, 90–91
watering gardens, 156, 163–65.
  See also drip irrigation
Waters, Alice, 319
weather, 78
  winterizing in cold, 105
websites, 300–301
  resources, 332–33
  seed, 196, 333
weed suppression, cover crops
  for, 180
weeding, 13, 77
wells, 54, 90–94
  disinfecting, 93–94
  for garden, 156, 163–64

maintenance chores, 91–92
  testing pressure tank, 92
  testing water, 91–92, 94
  water table and, 90–91
wheel bugs, 218
wheel hoes, 138
wheel seeders, 207, 210
wholesale (wholesalers), 312,
  313
Williamson, Rachel, 71, 184–87
winterizing, 105
wire weeders, 137
Wise Bauer, Susan, 223, 319,
  321, 323, 328
Wiswall, Richard, 287
witchers (dowsers), 45
woman farmer, use of term, 12
Women's Land Army, 20, 41
wood. *See* firewood
wood blocks, 151
wood chopping, 97, 107
wood splitters, 115, 118, 147
wood-burning stoves, 96–100,
  101, 104
  buying considerations,
    96–98
  cleaning out, 100, 102–3
  types of wood, 99
work expectations, for interns,
  315–16
work gloves, 83, 110, 118
workshops, 26, 87
World Wide Opportunities for
  Organic Farms (WWOOF),
  32, 33
woven wire fencing, 131
wrangling farm animals, 249–50
wrenches, 151

yoga, 74, 77–78
Young, Iris Marion, 79

zoning, 63, 66

ACKNOWLEDGMENTS

This book is a result of much combined knowledge, research, and shared experience. It is empowering to be able to do necessary tasks and work on one's own, but it truly takes a community to create a healthy farm.

I am deeply grateful to the women farmers (and their families) who generously shared their stories and welcomed me to their farms. Gail Hobbs-Page, Rachel Williamson, Erica Hellen, Krista Rahm, Susan Parks, Deborah Greig, Susan Wise Bauer, Elizabeth Blecha, Zan McKenna, Camey Stewart, Ashley Reece and Angel Shockley: I am inspired and awed by all of you! Thank you also to Temple Grandin who shared her valuable time and experience with a complete stranger in hopes that it would inspire other women in agriculture. Thank you to all of my friends and neighbors for all of your support, from helping to deal with jobs too big for one person, to offering the much appreciated glass of wine or beer, and really, just for being there.

My flower business would not be nearly as fulfilling and bountiful without the amazing wealth of knowledge I have been able to harvest from the members of the Association of Specialty Cut Flower Growers. Thank you for sharing your experience, expertise, questions, and humor. The support of my farmer's market customers and floral clients not only make it possible for me to continue to farm, but also provide the appreciation and wonder that keep me going in those hot, humid, weedy months of high summer.

This book was made possible by the determined effort of my husband, Michael. He showed me how to do many of the farm jobs portrayed in this book, as well as taking the photographs that really make this book a self-empowering manual. His unflagging support and willingness to read and offer critiques made sure I completed the book by deadline.

A huge thank you to Ann Treistman, my extraordinary editor, whose faith in this book made it happen and whose keen eye made this book into its best version. Thanks also to Sarah Bennett and Devorah Backman for their invaluable contributions to the book's success. And thank you to Nick Caruso for making the book come to life on the page with his beautiful design.

To my mother, Mary, who always tackles any job that comes to hand with energy and enthusiasm, thank you for passing some of that on to me. And to my sister, Elizabeth, who began a farming career long before I did. You are such a wonderful grower and my prime inspiration.

Of course, this book exists because of the many women farmers (gardeners, growers, teachers...) past, present, and future. This book grew out of the desire to share the stories of these women and encourage anyone that wants to grow healthy food in healthy soil, and to live a more sustainable life.